ADVANCED PRAISE FOR *TO SEE FAR*

Highly recommend! I found the book very hard to put down since it described so accurately what I remembered in my work at NASA's Johnson Space Center during the same time.

To See Far is an unexpectedly candid account of the technical, cultural, and human challenges encountered by Jim Van Laak, NASA's manager of operations during the assembly phase of the International Space Station (ISS).

Van Laak embraces the values of "tough but competent"* in his autobiographical account of the ISS program in its earliest phases. This was an era of big programs, big egos, and big changes in NASA, and it was ripe for internal conflict. The fact that NASA was forced to work with the Russians, heretofore archenemies during the Cold War, added yet another layer to NASA's internal conflict.

I was captivated by this deeply personal and candid account. Van Laak brilliantly captures the technical complexity of the ISS as well as his own "thrill of victory" at achieving major milestones of the ISS assembly phase. He also openly reveals his own "agony of defeat" at setbacks, letdowns, and betrayals both within NASA and with Russian partners in the program. He takes time to identify and explain his personal heroes, the individuals on both sides of the ocean who worked tirelessly for the success of the ISS program. He also identifies the various key leaders who did not rise above parochial perspectives or personal egos and, through action or inaction, made the challenge of ISS assembly even greater.

This account is not all "peaches and cream," but it does make the completion of the ISS assembly by an international partnership of former archenemies all the more remarkable, if not miraculous. And

Jim Van Laak deserves credit for his visionary, tough, and competent leadership during this phase.

Quote attributed to former NASA flight director, Gene Kranz, after the Apollo 1 tragedy that cost the lives of three astronauts: Gus Grissom, Ed White, and Roger Chaffee.

—Jon Hall, NASA (retired),
former project manager for Russian EVA hardware

"A captivating chronicle of US-Russian space cooperation."

To See Far is a definitive, firsthand account of the challenges, breakthroughs, and extraordinary teamwork that shaped US-Russian space cooperation from 1993 to the present. James Van Laak masterfully weaves together the tensions of the Cold War; the hard-won trust between NASA and Russian space engineers, managers, and crew members; and the operational triumphs that led to the construction and enduring success of the International Space Station. With rich storytelling and unparalleled insight, this book is an essential read for aerospace professionals, historians, policymakers, teachers, and students alike—offering a rare behind-the-scenes look at one of the greatest international engineering achievements in history.

—David M. Lengyel, PhD, NASA (retired)

To See Far: Conflict and Cooperation on the Space Frontier tells the story of how the US and Russia overcame tremendous odds to become partners in the International Space Station. It is a story with great meaning to me because I was intimately involved with both the conflict of the Cold War and the cooperation we shared in space.

As a US Marine Corps F-4 Phantom pilot, I trained to fight the Soviet Union, and while serving in Vietnam, I knew that Russian MiG-21 pilots were training the North Vietnamese to fly against us. The idea that I would one day become friends with a Russian MiG-21 pilot would have seemed ludicrous. Yet, on March 14, 1995, I launched on a Russian Soyuz rocket to the Mir Space Station with Lt. Col. Volodya Dezhurov as my commander. Before becoming a cosmonaut, Volodya had been a MiG-21 pilot in the Russian Air Force.

It was a remarkable change. That I had become close to Volodya and flight engineer Gennady Strekalov was indisputable. There was no sense at any time that these men regarded me as anything other than a colleague and/or friend. It was as if there had never been a Cold War.

—**Norm Thagard,**
first American astronaut to visit Russian space station Mir

Too See Far recounts the tumultuous early years of the International Space Station, which I covered as CNN's space correspondent. From my vantage point, the ISS partnership seemed fragile, but until I read Jim Van Laak's candid insider's account, I did not realize the extraordinary human and technical struggle to hold the program together. *To See Far* is an essential and eye-opening account of one of NASA's great projects.

—**Miles O'Brien,**
independent journalist and filmmaker

TO SEE FAR

CONFLICT AND COOPERATION
ON THE SPACE FRONTIER

JAMES VAN LAAK

Ballast Books, LLC
www.ballastbooks.com

Copyright © 2025 by James Van Laak

ISBN:
Paperback: 978-1-966786-58-0
Hardcover: 978-1-966786-71-9
eBook: 978-1-966786-59-7

Printed in the United States of America

Published by Ballast Books
www.ballastbooks.com

For more information, bulk orders, appearances, or speaking requests, please email: info@ballastbooks.com

To my brother John,
who should have been a part of this story.

TABLE OF CONTENTS

PREFACE

By Doris Hamill

Why should we go into space with so many problems to be solved here on Earth? Those of us who have devoted ourselves to the conquest of space have found our own answers that bring meaning to our daily lives, but our private satisfaction is difficult to translate into a motivation for public support. Similarly, we are hard-pressed to argue that the innumerable scientific discoveries and technology spin-offs space exploration has brought justify the endeavor. With no better argument in play, people often fall back on the mountaineer George Mallory's shrug: "Because it's there."

But perhaps both proponents and detractors alike are too close to the question to see the answer clearly. I think that if you were to describe the space endeavor to a bushman on the African veld and ask him why we should go, he would immediately answer, "To see far." He would know, as we all do, that the long view gives us a new perspective on reality that, even if not immediately useful, brings a richer sense of place and purpose.

We go into space to see far.

In the most literal sense, the vantage of space has allowed us to understand how finite and fragile our world really is. Apollo 8's stunning picture of Earth rising over the moon forced us to pause in turbulent times to consider how inextricably united we are, how petty our quarrels are on the cosmic scale, how beautiful our home planet is. To see a globe of white cloud and blue ocean and green and

i

beige continents inspires our awe at the natural perfection around us. Scenes of the night Earth sliding under a low orbit—with splotches of city light sparkling in the velvet blackness or glowing through cloud while lightning sparks perfectly white and silver-green auroras shimmer above the poles—bring an understanding of the balance between human greatness and human smallness that is simply not available from shorter views.

While these literal long views provide excellent reasons for having gone into space, they don't justify the continuing costs and dangers of the enterprise. Yet space, by its nature, provides other, more metaphorical long views that continually add a rich and fertile perspective to many human enterprises.

The scientific knowledge gained from the space perspective has utterly overthrown our parochial mythologies about the universe and our place in it. Our orbiting telescopes have let us look to the edge of time and space. Our scoops of moon rock have told us the story of Earth's birth. Our robots, grinding across the dusty gravel of Mars, swooping into the frost geysers of Enceladus, making radar images of the veiled face of Venus, and watching the sun fling out enormous chunks of its own flesh, have offered new and wondrous visions of reality that lengthen our imaginations and deepen our awe. Seeing far bundles knowledge into wisdom that adjusts our perspective on everything else.

The engineering required for the space endeavor, especially human spaceflight, has shown us the ragged edge of possibility and goaded us to exceed it. Every system we send into space faces critical struggles to first claw its way out of Earth's deep gravity well, then to hurl itself onward fast enough that it won't fall back into it. In these struggles, mass is the implacable enemy, and energy is the dangerous ally. Every gram of mass must justify the cost of its inclusion— "buy its way on," as we say. Yet every erg of energy increases the chance of catastrophe. Terrestrial engineering can always add more structure to protect against rogue energy, but adding structure to spacecraft calls for yet more energy to move it, and so the danger spirals upward. Spacecraft engineering must understand how each erg of energy affects each gram of mass. It has developed wholly new tools to wield that understanding, making computer models as indispensable to modern engineering as the lathe and the brake. These deep insights have shown engineers in every discipline new vistas in the realm of the possible.

But these new vistas also describe the limits beyond which conventional engineering cannot go. The difficult problems posed by space endeavors call for technology innovations to defy conventional limits. Space technologists stand on the shoulders of both scientists and engineers to stretch the long view just a little further, squinting uncertainly at possibilities that may take years, decades, or even generations to become reality. This goading by difficult problems directs our vision to fertile horizons of possibility and entices the patience and diligence needed to achieve them.

The global perspective, scientific knowledge, engineering understanding, and technology innovations are all long views worth having. But beyond them all, and providing vistas perhaps more important for the long-range future of humanity, are the challenges of cooperation that spaceflight demands. The complexity and criticality of spaceflight systems absolutely require support from large teams of people interacting seamlessly. The iconic view of a mission control center with its terraced consoles represents only the thin leading edge of the immense operations team needed to ensure the success of the mission, as this present work will illustrate. Each console has a "back room" of specialists digesting and reacting to detailed technical information. Before any operation begins, it is meticulously planned. Its envisioned operations are simulated and trained to prepare for every conceivable contingency, with responses to each foreseen "off-nominal" situation agreed upon and scripted in advance. When the operation ends, its interesting features are digested to extract any "lessons learned" to continually improve group understanding.

The technical challenges of spaceflight seem straightforward compared to the problems of integrating the team. As with any human endeavor, people come with their own abilities and shortcomings, their own passions and blind spots, their own egos and insecurities. The challenge of space operations—and the long view that spaceflight offers in the realm of human cooperation—is ensuring critical flight operations despite human foibles. Certainly, many other endeavors require large teams to cooperate smoothly. An airline, a skyscraper construction project, and an automobile factory must also coordinate the work of thousands of people. Space endeavors differ only because the tolerance for shortcomings is so low. Once a rocket starts generating a million pounds of thrust, once a returning spacecraft hits the top of an atmosphere at over seventeen thousand miles per hour, or once a probe descends to a surface thirty light-minutes away, everything must be within a small distance of perfection. There is no option to miss a departure time, reorganize work schedules, or bring the line to a halt.

The challenges of massive cooperation in space endeavors are offset by a few distinct advantages. Everyone perfectly understands the goal—mission success—and is entirely committed to achieving it. When disagreements arise on exactly how to do so, everyone understands that correct answers don't yield to ego, philosophy, or assumption but depend entirely on the laws of cause and effect. Right and wrong are adjudicated by mission success or failure, and there can be no appeal from the judgments of nature's laws.

These verities of spaceflight do not respect differences in culture, motivation, or philosophy. They apply to democratic capitalists as well as totalitarian communists, to government enterprises as well as commercial ones, and to atheists as well as fundamentalists. In this sense, the space endeavor provides an even longer view of cooperation among people—a prototype for cooperation across cultures, motivations, and philosophies.

This present work describes the design, construction, and shakedown of this far-seeing prototype. In the 1990s, the world's space programs undertook the challenge of building cooperation among peoples who approached the endeavor from utterly different cultures, perspectives, and motivations. It presented them with one of the most difficult technical challenges humanity had yet embraced and made the consequences of failure dire. Against all odds, the teams discovered that the verities of spaceflight provided the bridge that could span their differences to create durable mission success along with genuine respect despite deep differences.

The world beyond the space community does not think of itself as being driven by the same explicit motivations as spaceflight or see itself reacting so strictly to the laws of cause and effect. But perhaps this reflects a shorter view of the human enterprise. Perhaps the day will come when our common understanding of human purpose allows us to agree upon goals, when we have found laws of cause and effect in other human endeavors. On that day, we may acknowledge the prototype for cooperation pioneered by our recent space programs and be glad that those who sustained the enterprise across the decades had the vision to see so far into humanity's future.

1. PRELUDE

Loring AFB, Maine, 1982

Winter in northern Maine lets you know that you are an intruder in a profoundly hostile world. Its piercing beauty draws those hearty souls who relish icy winds, hard, frozen surfaces, and deep black nights, but it shows no mercy to the weak and unprepared.

This was the land my fellow fighter pilots and I endured, one week out of every five, as we sat alert in defense of our nation. On winter nights such as these, ours was a stark world—hard, cold, dark, solitary—where we served on the front lines of the Cold War. We felt our role acutely as lonely soldiers perched on the frontier, protecting our homeland from those who would destroy it.

Loring Air Force Base was home to a wing of B-52 bombers and air refueling tankers that were part of the United States' strategic deterrent. The bombers and tankers stood ready to launch an attack within fifteen minutes of being called. Our two air defense fighters were there to protect them.

While the job was deadly serious, most of our time was spent fighting boredom. Some studied to pass the time, while others read. I often ruminated on the circumstances that had brought us there and what they said about our world.

The view this evening was particularly poignant, dominated by a row of B-52 bombers braving the icy night. They stood guard, as they always did, ready to launch with their deadly cargo the instant an incoming threat was detected. But instead of being parked in their

1

normal roost at the north end of the runway, where their crews huddled in their earth-covered shelter, they were lined up near my post at the south end of the base. This put them closer to the active runway demanded by the unusually strong north wind.

Their urgency was heightened further by the presence of Soviet submarines close to the US coast. At such close range, their missiles were mere minutes from where we sat. To counter this, a pilot was stationed in each of the bombers, prepared to begin the start sequence while the rest of the crew raced to join them. Seconds counted; if war came, most of the bombers would be vaporized while still on the ground. It was a grim business.

The pale glow of a reading light inside the nearest bomber bore silent testimony to the lonely vigil of a young man fighting his personal battle to stay warm and alert within its frigid shell. His only company was the guard standing outside, fighting pathetically against the wind, there to ensure no enemy agents could choose this frightful night to sabotage our defenses. These unknown heroes bore the full force of the night against their many layers of clothing.

I shivered as I sipped my tea and pulled my jacket close, wondering again at the men and women who bore such duty alone and without glory. I wondered, too, at what had brought us to this point, both as a nation and as a species. How could it be that we would squander our treasure and youth to be certain of destroying each other? What kind of insanity had led our countries to put the future of humankind at such grave risk? The firepower stored on this base alone could destroy a dozen countries, and the combined weapons of the two superpowers could reset Earth's evolutionary clock by millions of years. Why were we doing this, and more importantly, what could be done to stop it?

During that winter of the Cold War, living on the tip of the spear, it was hard to be optimistic about the future of a world where so much was controlled by fear and hate, with millions of our fellows dedicated to our mutual destruction. How could we ever back down from this confrontation? How could we ever hope to overcome the fear and mistrust that had brought us to the brink of global nuclear war?

And if we tried—if we and the Soviets somehow decided that working together peacefully was better than threatening to destroy one another—could we learn to trust each other? Could we put aside decades of hating, fearing, training, and propaganda? Would any of us ever be able to call the other "friend"?

NASA Johnson Space Center, Houston, Texas, October 1993

We quietly filed into the conference room and took our places at the table. The room had been the venue for some of the greatest decisions in NASA's history, and we hoped that today would bring one of equal portent. The dark wooden table and walls that had borne witness to so much history now gave our efforts a somber air; my colleagues were unusually quiet as we assembled. Across the table, our Russian counterparts waited in grim silence.

On this day, the room held the leadership of the new international program to build a space station. Built on the bones of NASA's recently canceled Space Station Freedom, and hoping to take advantage of the long Russian history of space stations, this program was created to bring our two space programs together.

As obvious as it was, it was not our idea. President Clinton had canceled Freedom, making it clear that he would only support the billions of tax dollars it would take to build a new station if the program contributed more than the distant promise of science and technology. His support required that it contribute to urgent foreign policy goals as well, and we hoped that healing the scars of the Cold War would be such a worthy goal.

At the head of the table sat Dan Goldin, the fiery NASA administrator, and Yuri Koptev, his bellicose counterpart. Their grim determination flooded the room as their teams assembled for this important discussion. My colleagues and I would present our new plans to combine Russian and US assets into one unified space station in the hope of gaining their approval to proceed. Without it, the White House had made clear the space station was dead, and without a space station to build, the Space Shuttle program would not last long. We did not want to think about NASA's future without a human space program, and we knew the Russians faced a similarly bleak future.

The long conference table split the room, creating a sense of opposition. A half dozen NASA managers faced a similar number of Russians, with a handful of support people located behind us against the wall. Behind our leaders sat the interpreters with headsets to hear the speakers and microphones to transmit their translations, all of it recorded for posterity.

At the front of the long room, two screens displayed the briefing charts that would illustrate our plans. The charts on our side were in English, while the others showed what we hoped was the same information in Russian.

Dave Mobley stood at the head of the table, displaying a remarkable calm given the occasion. Tall, thin, and taciturn, Dave came from the Marshall Space Flight Center in Huntsville, Alabama. A well-respected veteran of the Apollo and Space Shuttle programs, Dave had spent months traveling constantly to find and resolve technical issues that could threaten the proposal. Now, he would present our plans and answer questions.

The meeting began with Dave's description of the US portion of the program. He carefully described the hardware we planned to use from the now-defunct Freedom design, including revised versions of the habitation modules, power systems, and structural trusses that would tie them all together. He briefly outlined their capabilities and the reasons they were included.

Next, Dave presented our ideas for introducing Russian elements in place of the Freedom elements that had been deleted. Their functional cargo block module would provide power and control for the early assembly phase until more capable elements could be added.

Their service module would provide early housing and support to the crew. Dave described a station that would be a genuinely joint effort, with each side contributing unique and necessary elements. If successful, the product would be a truly amazing machine, offering far more than either of our programs could alone. Our pieces would launch in the shuttle, theirs on Russian rockets, and join together on orbit.

The initial discussions went well, with only sporadic interruptions from the Russians as they corrected minor errors in our descriptions of their elements. The air was slowly warming; engineers live to discuss hardware and design, and the room was soon abuzz with technical details.

The calm was suddenly broken when Dave started to discuss how the pieces would be assembled in space. Because the majority of the assembly would require the space shuttle to be present, he focused on the steps to be accomplished during their many missions.

Without warning, sharp words began to fly between the Russians, leaving us to imagine what gaffe we had committed. At first, their exchanges were brief and sporadic, ignored by the translators, but soon they came rapidly, with conversations overlapping one another. The translators could not keep up, and we wondered at the tension building on the other side of the table.

Our new colleagues were clearly unhappy about something, but we had no idea what. After a few minutes, Koptev halted their argument and got up to join Dave at the front of the room. The moment loomed large as the powerfully built Russian began to duel with the slightly built American. My anxiety grew as Koptev jabbed at Dave with his wooden pointer while Dave stood his ground, his own pointer held tightly across his chest.

Once again, the Russian side exploded with a loud and rapid argument, far too fast for the interpreters to keep up. Returning to stand by his seat, Koptev raised his hand and asked for the interpretation to stop so they could caucus among themselves. Goldin assented, and we Americans sat in silence as the Russians argued back and forth with great bursts of noisy emotion. Clearly, this was no minor disagreement; the passion of their arguments washed over us.

One of the NASA safety people sitting behind us spoke fluent Russian. After a few minutes, he leaned forward and whispered to us while looking at the floor.

"They are unhappy about relying on the shuttle for assembly. They particularly don't want to fly cosmonauts on it," he said. "They consider it unreliable and much too dangerous to launch without an escape system."

This was a problem we had not anticipated. We Americans loved our shuttle and delighted in its beauty and grace, and we thought the Russians would be pleased by the opportunity to ride to space in such finery. Instead, they were angry to be told to accept additional risk for the sake of this program. In their eyes, our *Challenger* accident was proof that the design was flawed, and they did not want to risk the success of the program or the lives of their colleagues.

After perhaps five minutes of animated discussion, Koptev abruptly raised his hand to stop the Russian caucus. Sitting down next to Goldin, he had the look of a man doing something he hated but knew must be done.

Speaking through the interpreters, he told us that there were many issues and errors in our descriptions of their systems and operations but that what we were proposing was feasible. Then, he explained their concern regarding the safety of the shuttle. His scowl left no doubt about the seriousness of their concerns, and he made it clear that they would never voluntarily choose to fly a crew without an escape system.

But he went on to admit that without this new station, there might be no Russian space program at all. Despite the hallowed place their space adventures held in the minds and hearts of their citizens, the government's poverty had already reduced their once-proud program to a shadow of its former glory. Cooperating with NASA seemed like the only option to bring money and public support to their beloved space program.

Finally, he acknowledged that there was much they did not understand about the shuttle because it was the product of a different technical culture. He acknowledged that we must have good reasons for doing things as we did. He said that despite their deep concerns, if we were willing to risk our astronauts on the shuttle, they would consider allowing some cosmonauts to fly on it as well. They would seek approval from their government and, if acceptable to all, would begin to work with us on the new program.

Goldin quickly accepted this announcement with great enthusiasm, praising our new partners for their courage and foresight. He promised that we would listen carefully to their input and incorporate their wisdom into the new program.

The rest of us greeted the agreement with a mixture of relief and apprehension. Their agreement marked the culmination of months of feverish work and furious arguments, offering hope that we could move forward with our new program. But we were as conflicted as the Russians were about combining our program with that of our former enemy. While they talked about the safety risks they saw in the shuttle, we saw the many potential problems their participation would bring. Despite Goldin's words of shared victory, many of us saw it as an uneasy truce.

The meeting also reinforced my belief that solving the technical issues was the easy part. The far more intimidating step was committing to a long-term, high-risk project that demanded intimate cooperation among strangers. In this case, we were required to partner with those who, until only two years before, had been our mortal enemies.

Worst of all, we were making our commitment based on only a few months of high-level meetings punctuated with strong disagreements. Before us lay countless opportunities for our tenuous agreements to disintegrate as we went from high concept to concrete reality. And even if we overcame these challenges to work together beautifully, we wondered whether the Russian government and economy could support what they were proposing.

As we shook hands with our new partners, I was certain that every heart in the room was filled with a melancholy mixture of hope and concern. While we had passed a critical milestone, success lay far ahead beyond a veritable mountain of technical and human challenges.

I counted myself among the optimists who took comfort in knowing that we would scale that mountain together.

Mission Control Center, Houston, Texas, February 1, 2003

Nearly a decade had passed since we had joined forces with our Russian partners, and the intervening years had profoundly changed our world. Our once fiercely independent agencies had come together on a common task. Together, we had overcome technical hurdles, bridged communication gaps, and learned to trust one another.

NASA's proud shuttles had found steady work supporting our new space station. Russia's Soyuz and Progress spacecraft had transitioned from supporting their Mir Space Station to exclusively supporting the International Space Station. Mir, once the focus of controversy for its aged design and failing systems, was no more, having burned up in the atmosphere nearly two years prior.

I sat with my back against the wall in the management center of the mission control complex I knew so well. Although this was the same room where I had chaired over three hundred Mission Management Team (MMT) meetings to oversee the assembly of that station, today it felt like a very different place.

The room was filled with the muffled conversations of astronauts, engineers, and managers from the Space Shuttle program. However, unlike the typical NASA MMT, where the atmosphere vibrated with people working on issues and finding solutions, today the air was heavy, the mood somber.

We had gathered in the face of our worst nightmare—the disappearance of the space shuttle *Columbia*. Less than an hour before, the proud machine had been reentering at the end of a spectacularly successful science mission—one of the few that did not go to the station—when communications had stopped over northern Texas. The tragedy was confirmed a few minutes later when it failed to arrive at the Kennedy Space Center (KSC).

At home when the accident happened, I quickly raced to the control center to find the building filled with people in shock. At the head of the same table from which I had guided so many missions, senior managers methodically walked through our disaster response plan. They issued its somber instructions, frequently interrupted by reports of debris falling from the sky. Working through the thick plan, they named search and recovery teams, charged investigators and coordinators, and sent hundreds of people out for what would become months of gruesome work.

Although I had worked on dozens of space shuttle missions, I had not been part of this one. When it became clear that there was nothing I could contribute to their work, I quietly left and crossed the hall to the space station operations management center. Here, I found the man who had replaced me as head of ISS operations, Mike Suffredini, meeting with the station's engineering and operations team to review contingency plans for what everyone knew would be a long time without a space shuttle flight to ISS.

They discussed how much food, water, and propellant were on the station and how long they would last. Our careful plans ensured they could last for the better part of a year with no new supplies. Mike was assigning actions to people, and I kept quiet while he got them organized. When he was done, I spoke up.

"Have you talked about the Russian techniques for stretching the supplies?" I asked.

Mike looked surprised. "What do you mean?"

"During our time on Mir, the Russians showed us some innovative ways to get every last day out of the available supplies. They have had lots of practice doing that over the years, and you should have the techniques in your pocket just in case."

Most of Mike's involvement had been with NASA's hardware and operations, and it was obvious that he had not thought about what the Russians could contribute.

I continued, "Although we are far better off than the Mir was, it will be a long time before we get another shuttle up, and we should be ready to take advantage of every bit of capacity we have."

Mike nodded. "Good point. Please brief the team so they have all the options."

When the meeting broke up, I followed Mike to his office for a telecon with his Russian counterparts. They were well aware of the accident but unsure what it meant for the program. Mike carefully explained that we would be without shuttle flights for at least a year and probably much longer. During that time, we needed to operate the ISS using only Russian Soyuz and Progress vehicles to provide crew changes and logistical support. While it was a capability we had hoped not to need, the ability to support the station without the use of the space shuttle was one of the primary benefits of including Russia in the program.

The Russians knew the space shuttle would be grounded for an investigation, but they were surprised at the length of time Mike suggested. Whenever they had a failure, they too would stop operations until the investigation was concluded and recommendations made, but in Russia, this normally took a few months at most.

Mike explained that our technical and political system would not allow that and that we could be grounded for several years. He recounted that after the space shuttle *Challenger* exploded during launch in 1986, it was almost three years before the space shuttle flew again. Victor Blagov, the senior Russian flight director, quickly agreed to support us in any way they could.

Over the next few weeks, NASA and the world grieved for the loss of those seven astronauts, four of whom were close friends of mine. Once again, the dangers of flying in space were on display for all to see. By odd coincidence, it was the first shuttle flight in many years without a Russian cosmonaut on board. Even so, the Russians mourned the loss of the crew and the shuttle as fellow space explorers. They held services to honor their colleagues and set up a simple but respectful memorial in their mission control center.

It was a time of deep reflection for me. No longer buried by the day-to-day responsibilities of the ISS program, I had the leisure and perspective to marvel at the change that had taken place in just a few years. In less than two decades, I had gone from sitting on air defense alert, ready to fight a nuclear war against the Soviet Union, to working side by side with my former adversaries on a peaceful space station to grieving for our common loss. Our agencies had gone from fierce, isolated competitors to close allies. Along the way, we had learned to overcome decades of fear and mistrust to make our former enemies our partners, colleagues, and even friends.

This is the story of that transition, told through the events I witnessed and the people I came to know.

2. IN THE BEGINNING

Nothing about my early life hinted that I might someday be part of a grand enterprise. Indeed, everything suggested that I was destined for a very ordinary life, with nothing more to offer than an honest day's work. Yet, incredibly, circumstances would repeatedly align themselves in unexpected ways to give me a future far greater than I would have sought for myself. My only contribution was to pursue the opportunities when they arose.

In the 1950s, Schenectady, New York, was home to two major divisions of the General Electric Company. A heavy manufacturing plant downtown provided a solid foundation of traditional jobs that stabilized the local economy, while a nuclear research facility nearby drew a more elite workforce. Union College added Ivy League flavor, and the nearby cities of Albany and Troy brought the prestige of the state capital and Rensselaer Polytechnic Institute's technical academics. Set among the beautiful hills of upstate New York, it was a great place to grow up.

My life was busy and full, and I was a happy child. But beyond the happy distractions of youth, I often felt intimidated by the world around me. Many years later, I learned that this reflected an inherited low self-confidence and self-esteem. More than just painfully shy, I felt certain that I faced a very humble future.

The dramatic events of the time contributed to that sense of inferiority. Sputnik was launched before I reached kindergarten, triggering a wave of fear that our powerful adversary, the Soviet Union, was leaping ahead of the US. The government responded strongly, creating the National Aeronautics and Space Administration (NASA) and the Advanced Research Projects Agency (ARPA, now DARPA—the D standing for Defense) both in 1958, along with powerful educational initiatives designed to promote technical careers. Many of them would touch my life directly.

As I entered elementary school, NASA was beginning its grand adventure to put a US astronaut into space and, soon, on the moon. The new space program mesmerized the country, creating heroes, inspiring dreams, and guiding countless men and women into new careers. It promised great adventure.

My friends and I were among those drawn to this adventure, poring over *LIFE* magazine's gorgeous pictures and inventing games to share in the excitement. Yet, even as I embraced the space program, it made me feel tiny. Rather than being inspired to study and master its wonders, I felt certain that I did not have the intellectual capacity to participate. As much as I wished it could be otherwise, my relationship with NASA seemed destined for distant admiration.

This poisonous timidity hung over everything in my life. When I thought about a career, it was as a humble worker in the downtown manufacturing plant rather than as an "egghead" in the research facility. Hearing that an astronaut was coming to town, I could not imagine myself in the same room, let alone meeting them. The idea of becoming a colleague was ludicrous.

Fortunately, my low self-confidence was offset by a quick mind, and I soon developed a reputation as an intellectually capable, if lazy, student. Physical science and elementary mathematics were so obvious that one hearing locked the concepts in my brain with ease, leading to poor study habits that hurt me later. Machinery and electronics fascinated me, focusing my attention not only on how the bits and pieces worked together but also on the forces and reactions behind them. Yet, these portents for a career in engineering were stunted by my lack of confidence and ambition.

The seeds for change appeared when I came into contact with aviation. Although I had always watched the clouds and longed to be among them, aviation seemed to be just another dream that was out of reach—something that other better people did. When my dad would occasionally take us to the Albany Airport to watch the giant airliners come and go, I marveled at the steely-eyed professionals who flew those great machines—obviously a different breed from mere ground dwellers such as I.

I was sixteen when I got my first short ride in a light airplane. It came as a shock to see that the pilot was just a normal human being, not a sky god. The ride was magical, and the new perspective was awe-inspiring. Issues that seemed overwhelming while on the ground suddenly shrank into insignificance. When the pilot let me handle the controls for a few minutes, the machine did not tumble out of control but responded docilely to my gentle touch.

Those too-few minutes aloft changed my life forever, filling my head with dreams of a lifetime spent in the sky. With the support of my dear mother, I began flying lessons while a senior in high school and completed my license during my freshman year in college. The self-confidence and energy I found in aviation gave me a new outlook on life.

My dad, however, scoffed at the idea of an aviation career, calling pilots "glorified bus drivers." His disdain could not tarnish the magic that would forever change my life, but it taught me to keep my dreams to myself.

Something truly profound had happened. I had tasted flight and could see in the experience the basis for a lifetime of challenge and adventure. My lazy, under-motivated, and self-conscious self was overtaken by an irresistible love for the freedom and challenge of flight. If a future in flight meant studying and learning to make decisions, then those were obstacles I would face. If it required hard work and courage, then I would devote myself to their pursuit. I even began to think that maybe—just maybe—I could accomplish something meaningful along the way.

At about the same time that I found my passion in aviation, I discovered that I had an intuitive understanding of elementary physics. It did not just come easily; it seemed to already exist in my brain, ready to go to work. I took the Scholastic Aptitude Test for physics as a high school junior—a year early and with only a few weeks of studying under my belt—with the expectation that this would guide my learning and allow me to do well when I took it as a senior. To everyone's shock, I got a perfect score of 800, and my quiet life was suddenly burdened with other people's expectations.

Encouraged by my teachers and emboldened by my sudden success, I chose to pursue a degree in physics at the nearby Rensselaer Polytechnic Institute. While college demanded much more effort than I had ever given to high school, the physics classes were a breeze—even fun.

As I adjusted to college life, events far beyond my small world suddenly broke over me. Mine was the last class eligible to be drafted for the war in Vietnam, and the draft lottery that year paired my birthday with the number seven in priority for conscription. Like many, I had always considered the number seven to be lucky, making this seem like an especially cruel joke.

The low draft number guaranteed I would serve in the military, but while I was a patriot, the prospect of military service terrified me. The war in Vietnam had cast a dark pall over our society, undermining public confidence in our leaders and creating deep divisions among well-meaning people. Even though the war was beginning to wind down, fighting remained savage both in the fields of Vietnam and on the streets of home. The turmoil deeply confused and conflicted me, making me desperate to understand why I and others were being sent to kill people on the other side of the planet.

Physics did nothing to answer that question, so I decided to find something that would. Adding history, political science, psychology, and literature to my studies meant giving up physics as my major, but I accepted the cost in order to be the responsible citizen I wanted to be. My decision to major in economics would cause many awkward moments later in life, but I never regretted the decision to broaden my understanding of the world. The department chair also helped me continue my studies in science and engineering, giving me a well-rounded education.

My biggest worry was that I could not fulfill my responsibility as a soldier. The discipline, hard work, and courage that the military demanded were my weakest traits, and I worried that they would be my downfall. Considering all the options for fulfilling my service obligation, I was relieved to find that RPI had an Air Force Reserve Officer Training Corps (ROTC) detachment. The staff there promised to teach me the skills I would need to master the military culture and assured me that if I did well enough, I might even become an air force pilot.

The possibility of someday flying a graceful jet provided the motivation I needed. While flying light airplanes had already inspired me to work hard and provided some critically needed self-confidence, the dream of flying such a machine was irresistible. To seal the deal, the air force offered me a scholarship, which I gratefully accepted.

The future that had looked so grim just a few months earlier began to take on an optimistic hue. Even my father, encouraged by my air force scholarship, began to support my aviation interest, buying an old fabric-covered Taylorcraft airplane for me to gain experience. Eventually, he helped me move up to an only slightly newer but much more challenging Globe Swift. I flew this graceful machine at every opportunity, and it was literally my pride and joy.

But on July 29, 1973, during takeoff in the Swift, the engine caught fire and quit. I was forced to land on the very end of the runway with the landing gear up, damaging the beautiful plane. It was not until we stopped moving that I began to smell smoke. My passenger and I got out without injury, only to stand by helplessly as the plane burned to the ground.

The horror of watching my beautiful plane burn wrenched my heart, but it was nothing compared to the humiliation and self-doubt that consumed me afterward. My father came to the airport to get me and lectured me all the way home, taking me to task for things he knew nothing about and leaving me feeling like the smallest person in the world.

The next day, the FAA came to investigate, adding the potential of being found in violation of some obscure regulation to my distress. After examining the wreckage, the inspector concluded that I was not at fault and even complimented my skill in handling the emergency. He determined that a fuel line had fractured due to metal fatigue, spraying the engine with raw fuel and causing the fire. He also pointed out that had I been much higher, the fire would have killed me before I could get to the ground. Luck was with me that day.

Despite the official vindication, the accident shattered the confidence and sense of identity I had found in aviation. My youthful conceit of invulnerability was lost forever; accidents had become real to me. I was consumed by a fear I had never known before—its icy shaft penetrated my heart, threatening to paralyze me and wipe away all that I had hoped for.

But as awful as it was, I managed to survive it all, both physically and emotionally. At the ripe old age of twenty, I got my first lesson in how adversity brings growth. Motivated by my deep love for aviation, I overcame my fear and began flying again within days, borrowing planes from sympathetic and supportive friends. Despite the ice in my stomach, I committed myself to learning all that I could about the art and machinery of aviation—what makes it work and especially what makes it fail. I also looked inward, becoming far more critical of my own decision-making, too often finding it sadly wanting. While the price of the accident was dear, the pain and changes it wrought would set the stage for all that I would become and contribute later in life.

This new commitment to learn as much as I could about aviation led me to spend my weekends and summers working at a local airport.

The airport manager and mechanic, known to all as Brownie, taught me about aircraft systems and how to diagnose problems, fix what was wrong, and test my fixes both on the ground and in the air. He insisted that I test-fly the product of my work, increasing my attention to detail and confidence in myself. When some of these test flights did not go well, they taught me to calmly deal with a crisis while analyzing the cause. This experience would also serve me well in the future.

Returning to RPI that fall brought me another of the great gifts of my life—getting to know Doris Hamill. Already a member of the Air Force ROTC, Doris would not only become my wife, but she would also become the most important influence in every dimension of my life. Her intelligence and energy challenged me, and we soon formed a strong platonic friendship. We spent countless hours exploring the issues and mysteries of life, sharing our experiences, questions, wisdom, and mistakes. We became soulmates long before romance entered our lives, so when we realized our lives were becoming intertwined, a profound love had already taken root.

We married shortly after graduation and were soon both on active duty in the air force. Doris worked as a research scientist at Brooks Air Force Base in San Antonio, Texas, while I went to pilot training at Laughlin Air Force Base in Del Rio, Texas, one hundred and fifty miles away. Doris flourished in the world of technology and science while I struggled with a toxic mix of rigid military culture and lingering self-doubt. Fortunately, Doris kept me on track long enough to find my way. With her love and support, I forged on and was amazed to graduate at the top of my class.

Graduating high in my class was essential to our future; it gave me priority in choosing the airplane I would fly and where I would be stationed, creating the opportunity for Doris and me to be stationed together. We chose Griffiss Air Force Base in Rome, New York, where I would fly the T-33 and F-106 while Doris worked at the Rome Air Development Center. She was soon deep into the development of advanced technologies for the Pentagon, while my duties took me all around the US and Canada, including many weeks spent on alert at our detachment located at Loring Air Force Base in Maine.

Winning my wings as an air force pilot was a proud moment, and it soon led to the next level of education I so desperately needed. As a brand-new fighter pilot surrounded by cocky colleagues with far more experience and bravado, I struggled to find my place. I soon discovered the answer lay in a delicate balance between the confidence needed to take on challenging tasks and the wholesome humility required to make wise decisions.

It was a difficult lesson. Flying a single-seat fighter taught me that I could do things I would never have imagined possible, like flying formation in turbulent clouds at night with only a dim light for reference, overcoming mortal fear as I worked to expand the ragged edge of my ability. When a friend or colleague was killed, it reminded me how close I had been to disaster on more than one occasion. The echo of the fear from my accident remained, but instead of paralyzing me, it forced me to be brutally honest about my abilities.

The air force helped me grow in other ways as well. My civilian maintenance skills led to several assignments in the squadron maintenance organization. One put me in charge of over two hundred young men and women who kept our airplanes flying in all kinds of weather. This gave me a crash course in leadership and a deep appreciation for those whose thankless work makes great things happen. Another assignment put me in charge of the quality assurance team, during which I flew over a hundred maintenance test flights. I loved testing these sophisticated machines, further advancing my understanding of their systems and my ability to function under pressure.

Although I loved the airplanes and people, I was not cut out for the military. In 1984, not long after my tour in the air force was up, Doris was offered a job as a civilian program manager at the premier technology arm of the military, the Defense Advanced Research Projects Agency, or DARPA. It was a great opportunity for her, but it was located in the heart of the Washington, DC, metropolitan area.

Moving was difficult—giving up both our quiet home in upstate New York and my easy access to flying—but it was clearly the right thing to do.

My transition to life in the Washington, DC, area left me deeply unhappy and, I admit, unpleasant to be around. Months of unemployment, punctuated with rejections from hiring managers, were devastating, creating a profound sense of worthlessness. Once again, I feared that I would end up trapped in a mediocre job in a nameless organization. When I was offered a job as an analyst on a classified program for a different part of DARPA, I reported to work with the dread of someone destined to cut stones with a sledge.

To my amazement, the job turned out to be fascinating and challenging in ways I had never imagined. Suddenly, I was immersed in technical issues that had no apparent solution, challenged to use my brain as never before. Working with the finest academic institutions in the country—MIT, Stanford, Georgia Tech, and others—I was stunned to be treated with the same respect as my more learned colleagues. As a valued team member, my talent for understanding complex systems blossomed, and I learned and grew at a phenomenal pace.

As I settled into my work at DARPA, my weekend flying introduced me to an executive in the Space Shuttle Program at NASA Headquarters. Jon Michael (Mike) Smith soon became a close friend and confidant. His enthusiasm for NASA, although tried by the aftermath of the space shuttle *Challenger* accident, was inescapable. He was friendly, intense, enormously intelligent, and overflowing with energy.

Mike owned a small biplane called a Pitts Special that he used for aerobatic flying and the occasional airshow. He loved his Pitts, but its engine had a disconcerting habit of cutting out during critical maneuvers. Mike had spent thousands of dollars on specialists, but the problem remained, and his frustration was palpable. I wanted to help and felt that I could. My time with Brownie had given me the skills to obtain an FAA license to work on aircraft, and I was even building one in my suburban garage in Alexandria. When I casually offered to try to fix his problem, Mike was amused but let me give it a shot. Two weeks later, I pronounced the problem solved. After test-flying the Pitts, Mike was ecstatic.

Seeing potential in my work, he started looking for a way to bring me into NASA. He shopped my name around the headquarters building, and I interviewed with a number of offices. Soon, I got a call from Dr. Michael Greenfield in the safety organization. As a result of the *Challenger* accident, which had taken the lives of seven astronauts, including teacher Christa McAuliffe, Michael was tasked with creating a new office to do independent technical assessments of important issues. He invited me for an interview.

Michael was hard to miss, with his jet-black hair, full beard, bright eyes, and garrulous manner. He chattered excitedly about the potential for his new office but grumbled over the team he had inherited. He wanted people with deep curiosity, good analytical skills, and operational experience to ask the right questions and pursue solid answers. When he made me an offer, I took it.

As 1987 came to a close, I began my amazing adventures working for NASA, the agency I had revered since childhood. This first, modest step—a small job in the headquarters safety organization—came at the best possible time and place. The accident had shaken NASA to its core, and it was hard at work reviewing the entire program from the bottom up, covering everything from the engines to the toilet. Michael encouraged me to take advantage of every opportunity to learn, and I did.

My job doing independent technical assessments exposed me to an enormous range of topics that needed to be resolved in order to return the Shuttle to flight. I was constantly challenged to learn new systems and disciplines, an imposing task made possible with the generous help of agency experts—from engineers to astronauts to flight directors to international authorities in arcane technical disciplines. This also allowed me to develop techniques to accurately assess the robustness of designs and the wisdom of proposed plans. It was both enormously confidence-inspiring and profoundly humbling—another combination that would serve me well in the future.

As much as I enjoyed that job, I also hoped it would be a step toward something greater. Even more than at DARPA, I was driven to use every bit of my intelligence and talent in my work. My total commitment earned me a reputation for honesty and thoroughness—two qualities highly valued in the NASA community—laying the foundation for all that followed.

Michael Greenfield, eager to keep me on his team, promoted me as quickly as possible to the maximum GS-15 level. But despite my success at NASA Headquarters, I longed to be at a field center, getting my hands dirty with the real work of spaceflight. I wanted to contribute, not just critique.

After two years of working for Michael, I began seriously searching for a new job in the field. Doris had completed her tour at DARPA and was willing to go wherever the job might take me. While I remained open to all possibilities, one destination stood out.

The Johnson Space Center in Houston, Texas.

3. WELCOME TO HOUSTON

The bright morning in September of 1989 showed no hint of relief from the brutal summers that grace eastern Texas. Especially for someone who grew up in a much colder climate, visiting NASA's Johnson Space Center (JSC) and tolerating the high temperatures and near-saturation humidity was always a challenge. In addition to the heat and humidity, the air that morning reeked of the sickly-sweet smell of the chemical plants in Pasadena, ten miles to the northeast. The unpleasant smell penetrated even my hotel room and cast a slight pall over the coming day. I tried not to think of it as an inauspicious start to what I hoped would be a turning point in my career. I had come to this hot, humid, flat, and featureless land in search of a more meaningful job, and this was the day of my most promising interview.

Webster, a town halfway between Houston and Galveston, Texas, was filled with small engineering firms and strip malls. Remarkable only because it lay just outside the gates of JSC, it had been the home base for NASA's human spaceflight program since the mid-1960s. Best known as the home of the astronauts and the flight controllers who populate mission control, it had been the focus of popular and media attention for decades. Anyone serious about human spaceflight in America would find their way to JSC.

The moment I walked into the Eagle Building, located about a mile from JSC, my mood and outlook improved dramatically. Dennis Webb, the newest and youngest division chief in JSC's Mission Operations Directorate (MOD), welcomed me into his office with an air of boyish, utterly unselfconscious enthusiasm. Dennis had been told about my work and thought I could add some of my unique expertise to his team. He presided over a motley mix of engineers who were hard at work becoming experts in the systems and operations of the planned Space Station Freedom. His responsibilities included the on-orbit logistics and maintenance for the space station, and he thought my qualifications made me an ideal candidate to lead the work.

President Ronald Reagan had proposed the Space Station Freedom program during his State of the Union address in 1984. Envisioned as a bold step toward expanding human presence across our solar system, the program had suffered from both technical and management problems. These issues led to a series of redesigns meant to meet cost and value goals. The program's troubled history gave me some pause about joining the effort, but Dennis radiated confidence that the program was finally moving forward.

Dennis's office in the northwest corner of the Eagle Building was typical for NASA engineers and managers. Unlike the people I had worked with at DARPA, whose sterile offices hid classified materials in locked cabinets, Dennis and his colleagues lived among tables piled high with drawings and thick technical documents. At the time, Dennis's team was located in one of the many buildings around JSC that provided overflow offices for the US space program. It was natural to wonder whether Dennis's off-site location meant the work was not considered important, but he assured me they would be moving on-site within a few months. I gladly pushed that concern aside, giving him my full attention.

Dennis and his team radiated energy and optimism. Many, including Dennis, were veterans of the Space Shuttle program side of MOD, drawn to the new program where the challenges were still fresh. His job was to shape them into the experts who would operate the space station during its assembly and throughout its lifetime. He was absolutely bubbling with delight about the opportunity he had been given.

As part of my interview, Dennis brought in a group of young and mid-career engineers who made up the core of the logistics and maintenance team. Kevin Watson and his colleagues knew my qualifications and were eager to talk with me about their challenges and to learn what I thought.

"Our job is to plan logistics and maintenance to maximize productivity and minimize failures," Kevin said. "The station is being built to do research, and we need to minimize the burden of maintaining it."

The team was tightly focused on responding to a report known as the Fisher-Price Study. A group of engineers and operators had been chartered to evaluate the maintenance burden that the space station could demand. The report suggested that if not correctly managed, maintenance could make research extremely difficult and introduce significant risks to the program. Congress was alarmed by the prospect of spending so much and getting so little return, making it an urgent issue.

I had read the report before coming and thought I understood the concerns. After listening carefully to their views, I offered what I hoped was a clear vision of the path ahead. Their smiles reassured me that we were all on the same page and that they were eager for my help.

The group was particularly excited by the fact that I had personal experience with maintenance and logistics. My work maintaining air force aircraft and my FAA license to work on airplanes gave me a great deal of credibility. Approving looks shot around the table when I talked about having modified several airplanes on my own and even having built one from scratch. Not many senior managers knew what it

The Fisher-Price Study

In 1990, a team led by astronaut Bill Fisher and robotics expert Charlie Price estimated it could take up to one spacewalk every workday to keep the station functional. This prediction of 3,200 hours of extravehicular activity (EVA) each year would cripple the scientific value of the station and greatly upset congressional leaders.

was like to build flight hardware, let alone to put one's own tender body on the line to test-fly it. The team appreciated that such an experience had brought me a more profound commitment to detail than most managers had, which they desperately needed.

They went on to describe three principal issues they hoped I could help with. The first was the Fisher-Price Study, suggesting the station could demand a great deal of astronaut and ground time to keep everything operating. In addition to being an enormous drain on resources meant for research, sending crews outside the spacecraft five times a week posed a significant safety risk to the program. The team's first priority was to reduce this burden to something more manageable.

The second issue was tightly tied to the first. At the time, program planning had not allocated any crew time or material resources for maintenance or logistics during the assembly phase. Senior managers believed their expensive new hardware should last at least five years without needing replacement. No amount of arguing could convince them that hardware sometimes failed early, a phenomenon called "infant mortality," and that they needed to prepare for it.

While I was following these points eagerly, the conversation began to resonate in a different part of my brain. The independent assessments I had done at headquarters had taught me to think about the overall effort, not just one aspect of it. Although I knew the maintenance issue was critical to the success of the program, I also understood management's decision to balance it against what they saw as higher-priority challenges. If I took this job, I would make a point to understand those other challenges and work on making the entire program successful, not just the maintenance piece. I would not seek parochial goals over the success of the program.

The third problem was related to the first two. MOD and the program needed a leader who could command the attention of senior managers.

"Our voices are spread out across many organizations with no unifying vision. We need a senior leader who can speak for the community," Kevin said.

This was why I was being received so enthusiastically. Beyond having the technical skills and personal experience, they needed a high-grade manager who could be taken seriously by program leaders. Most people at centers like JSC looked down their noses at headquarters people and their inflated grade levels, but in this case, they saw my GS-15 as an asset!

Dennis and his team had even come up with a homey title and acronym for the position I was to fill—the maintenance operations manager, or MOM. It gave a warm and personal touch to the position, and their attitude toward it was everything I could hope for. Such enthusiasm and personal attention were intoxicating—they needed and wanted me!

Another plus for the job was the fact that it was located in MOD under the leadership of the famous Gene Kranz. Although this was a minor role in just one of a dozen divisions under his control, the thought of working in his organization was a thrill that kept me interested, even though the work itself sounded rather pedestrian.

As the day of my interview wore on, Dennis could tell that even though the meetings were going well, I was still undecided. In the early afternoon, he took me into his office and asked what I thought.

"You have a great group, and I really love their intelligence and enthusiasm. It is an area where I think I could make a real contribution. But to be honest, I am not sure it is right for me."

Dennis gave out a little laugh as if he had expected exactly that answer. Then he said, "Not to worry, I have a secret weapon."

We were soon in his car driving to JSC. After parking in front of Building 1, the impressive nine-story headquarters building at the front of the campus, we took the elevator to the eighth floor and rounded the corner to the main office for MOD. The secretary greeted Dennis warmly and told us to have a seat. "Gene will be with you shortly." It suddenly dawned on me that Dennis was bringing me to meet Gene Kranz himself!

My three years working at NASA Headquarters had led me to believe that Gene was something special, even in a community of extraordinary people. One of the early flight directors in the US space program, his contributions were clearly visible in the success of Apollo. His two most famous moments came during Apollo 11 when he led the first manned landing on the moon and during Apollo 13 when he contributed to the rescue of the crew after the spacecraft suffered a critical failure on the way to the moon. Although both were huge team efforts, his personal leadership played a critical role in their success. I had been thrilled at the prospect of working in Gene's organization, thinking that perhaps someday I would get to meet the great man. Now, I found myself about to be presented to him for inspection. If I had known ahead of time, I would have been scared silly. Instead, I struggled to keep my composure and look professional.

We were ushered into Gene's office to sit at a circular table by the front windows. Gene welcomed me warmly and introduced me to some of his senior leaders. Then, he got down to business.

"So, what exactly is it that makes you so doggone special that I need to hire you?" he asked. It was less a question than a probe to measure me as a man. His direct approach caught me off guard, and I nearly blew the interview.

"Well, I don't claim to be anything special, sir." Despite my efforts to be casual, I could not shake the reality that I was talking to The Man. "In fact, I am really an expert in nothing at all. My talent is understanding the big picture, so I focus on that rather than trying to become a super expert in any one discipline. I have spent my life flying, designing, and working on sophisticated aerospace systems, learning everything I could from people smarter than me. And even though I am not an expert in any one thing, I try to know a useful amount about nearly everything. From what Dennis has told me, it sounds like that is what is needed here. You need someone who can pull the big picture of logistics and maintenance together, present it in the context of the larger program, and guide the team to the right answers. I think I could do that well."

I would later wonder at my calm in the presence of The Man, but at the moment, it was easy. Despite the riot churning in my gut, I held his gaze, did my best to answer his questions, and acted like a trusted colleague.

Over the next thirty minutes, Gene continued to probe gently but deeply, demonstrating the leadership that had made him famous. He listened closely, and when he spoke, it was with authority and plain language, a combination serious people appreciate. I did my best to show that I understood and respected the work and would do the best job I possibly could.

Toward the end of the interview, he proudly told me about his organization and its special role, extending an invitation to come join his team. With that, I knew I had passed his muster and fought to hide the excitement I felt rising inside. He ended the interview by extending his hand and saying, "Come join us, Jim. We'll do great things. Your country needs you, son."

Dennis beamed as we left, knowing Gene had made the desired impression. As we drove back to the Eagle Building, we talked about what it would take to complete my transfer. Although I did not fool him, I made a point of not accepting the job outright, telling him I needed to discuss it with my wife that evening. When we shook hands at the end of the day, he knew he'd found his man.

I called home to tell Doris about my day. Even though I tried not to show my excitement, she knew that I had found my place and that this was what I needed to do. I promised her that I would sleep on it and treat it as the serious decision it was. We would discuss it once again before I would give Dennis my answer.

The night passed sleeplessly until the new day dawned.

4. INTO THE FRAY

The first week of January 1991 dawned clear and bright, its gentle cool providing a welcome relief from the blistering heat that had dominated the year. Our new home was located on the runway of an airport community that brought aviation close to home, and my exciting new job promised plenty of challenges and excitement in the months and years ahead. Optimism was easy as I reported to work in the Eagle Building.

Dennis Webb began by introducing me to my MOD colleagues, and Kevin Watson took me to meet some of the space station people with whom we would be working. They made up a heady mix of individuals who had built and operated the spaceships I had watched from afar. Now, they were welcoming me as a new colleague. It was a dream come true.

The Johnson Space Center's role in human space operations had been created from whole cloth by some of the very people I was meeting. The Mission Operations Directorate was here, of course, with its cavernous control center, razor-sharp flight controllers, and superb training team. The Space and Life Sciences Directorate oversaw the health of the crew, performed research, and flew scientific payloads on the shuttle. The Flight Crew Operations Directorate managed the astronauts, the support teams, and a fine aircraft operations organization based at the nearby Ellington Field. The Engineering Directorate had deep expertise in every imaginable field and some of the finest test facilities in the world, including a site dedicated to hazardous operations located in White Sands, New Mexico. The Safety and Mission Assurance Directorate had engineers and specialists to support and monitor the activities of every aspect of the center. The work was supported by a first-class team of experts in human resources, finance, legal, security, and every other function required to make JSC an efficient and productive place. Although NASA had thousands of people performing critical functions around the country, there was no doubt that Houston was the center of the action for human flight operations.

It did not take long to see the issue that had brought me to Houston. While there were hundreds of talented and dedicated people involved in planning for the maintenance of Space Station Freedom, the work was not well-focused. Everyone was pursuing what they thought was the right thing to do with only the highest-level requirements to guide them.

Part of the problem was the way the program was structured, with strong organizations within the NASA centers and a relatively weak central program office located far from the action. After the 1986 *Challenger* accident, NASA had been directed to move the program office to Washington, while Congress insisted on giving their constituent centers as much authority—and bragging rights—as possible. Congress gave the program budget authority directly to the centers, greatly limiting the program manager's ability to control the work and its costs. The result was a program office with little control over the work.

Although the maintenance team had a boss in the program office, the individual who came closest to providing day-to-day leadership was Barry Boswell, the logistics and maintenance manager for the JSC-led portion of Freedom. Tall, lean, and friendly with a gentle but businesslike demeanor, Barry exuded the quiet confidence and competence common to most JSC leaders. I liked him immediately. I later learned that some of his unflappability came from his service as a riverboat commander during the Vietnam War.

Barry held one of the most important roles in my new world. Freedom consisted of work packages at NASA's major centers: the

Johnson Space Center in Houston, the Marshall Space Flight Center in Huntsville, Alabama, and the Lewis (now Glenn) Research Center in Cleveland, Ohio. Each center had managers responsible for ensuring their hardware was reliable and maintainable. Barry did this for JSC, but his role also encompassed much of the technical integration work for the whole of Freedom. His group planned and managed things like the tools and equipment for spacewalkers, the use of robotics to move large items around in space, and much of the logistics system. This gave JSC and its managers great responsibility. I found Barry to be both respectful of that responsibility and not shy in fulfilling it.

Kevin took me to meet Barry in his office in Building 1.

"Welcome, Jim." His friendly manner and firm handshake made it clear that I would enjoy working with him. "I am delighted to have you on board. Kevin has told me all about you, and as you will see, we will be working together a great deal." The twinkle in his eye offered both a welcome and a harbinger of challenges to come.

In addition to the logistics and maintenance for the hardware JSC was developing, Barry also managed much of the required infrastructure for the entire station. His influence touched every corner of the program, and whatever I hoped to accomplish would need to align with his needs and plans.

"I've been studying the Fisher-Price report," I said, "and it seems to be awfully conservative. It looks entirely possible to fix a lot just by cleaning up a few details."

"You are exactly right that it is conservative, but that is because it was always intended to be part one of a two-part process," Barry said. "Part one was to scope out how bad the problem could potentially be and get the attention of the folks who thought it inconsequential. Its prediction of five extravehicular activities (EVA, or spacewalks) a week definitely got the attention it needed! The second part would show how that conservative assessment could be whittled back to something we could live with. In fact, much of that work is already done."

"That makes sense, but if so, why is it so controversial? I believe I see real signs of concern, not just handwringing."

"Ah, now there's the rub," he said with a sly smile. "We know the basics of what needs to be done, but the devil is truly hiding in the details. More importantly, ours is not the only problem being worked on. Right now, the bigger issue is creating a design that can be built, and for many folks, maintenance is an afterthought."

My job was to keep maintenance from getting lost, particularly as the design changed. Even as we spoke, the Freedom program was in the midst of yet another redesign. From its inception, the program had been plagued with immature design guidance that made progress almost impossible. Such guidance, known as requirements, provides a precise definition of what the system is supposed to do and how. Without mature requirements, tremendous amounts of time and money could be wasted, as Freedom had already demonstrated. Designs and managers had been shuffled multiple times, and everyone hoped that this one would be the last.

"Will we have a voice in the redesign?" I asked.

"We have our chance to make inputs, but the design still has a long way to go to be a good spacecraft. We are just one part of the team."

Things were obviously happening fast, and I wondered if I would be able to get up to speed in time to contribute. I had barely been introduced and was already struggling.

Barry went on. "Maintaining a healthy station is the goal, but don't lose sight of the role of logistics. As I recently told my boss, John Aaron, while the maintenance issues appear workable, the logistics implications are *ominous*."

His emphasis on the last word was chilling.

"What does that mean?"

"It means every maintenance action will require an enormous amount of logistical support, from whole spare units to individual parts. They have to be bought, launched, and stored. If they go outside, they need to be EVA-compatible. If the robot is to handle it, it needs to be designed for that, with handles, temporary storage, and lots more. Every step adds a ton of complexity to an already-challenging problem. The decisions we make now will have an enormous impact on the ease and effectiveness of the maintenance program."

The mental picture I had created of my job and responsibility started to blur with layers of new details. In theory, my job was simply to lead MOD's efforts to perform maintenance on the station, but Dennis Webb and his team wanted me to bring added attention and focus to the work. While I wanted to do that, the more Barry described his work, the more unsure I became that I could make a difference.

I left our meeting with my head full of conflicting information and ideas. The air force had taught me a great deal about logistics, but this program took it to an entirely new domain. Things I had not thought of, like how to keep a spare part from floating away, were now critical elements that needed real solutions. Barry had plans that sounded good to me, but I knew I was way behind and in no position to judge. I committed myself to learning everything I could from him.

Once again, luck had put me in exactly the right place at the right moment. The latest redesign was just beginning, and much of its technical team was coming to JSC to participate. It was the perfect time and place to meet everyone and learn about the design and its issues. I hoped that if I got my act together quickly, I could even start to influence the redesign, but I knew that jumping in prematurely would destroy my personal credibility among those whose help and respect I most needed. A poorly chosen comment would either be ignored or interrupt the work of hundreds of engineers across the country. Taking my cue from Mark Twain's adage that it is "better to keep your mouth shut and be thought a fool than open it and remove all doubt," I resolved to hold my tongue unless I was sure I had a solid point to make. The pace of the work reinforced that wisdom.

The center of the redesign activity was Building 9 at JSC. This complex housed the full-scale mockups of the space shuttle used by engineers and favored as backdrops for television reporters. There were also mockups of the station modules that designers and operators used to get a sense of how their ideas might fit into the finished product. They created an important training venue for crews and others, providing just the right balance of activity and opportunity for quiet thought. I loved to walk the floor and, when the opportunity permitted, spend time inside the mockups. It was a great morale booster when our work bogged down and helped me learn more about the problems we had to solve.

The design meetings were filled with impatient energy, as most of the team had been working together for years, and everyone wanted this to be the last redesign. Every tick of the clock meant money being spent, and every significant design change set the program back months or even years. Although Congress had been patient so far, everyone knew it was past time to get on with it.

The challenge seemed overwhelming. The space station was far too large to be launched with one rocket, so it would have to be launched in pieces and assembled in space. However, designing a machine to be assembled in orbit was much harder than traditional engineering tasks for many reasons. The biggest challenge was that once the assembly began, we were committed to keeping the station operating continuously, no matter what. Although a spaceship in orbit looks peaceful, it is constantly performing many critical functions needed to keep itself alive. Our assembly plans had to keep all these functions operating without significant interruption, or all would be lost. There was no room for error.

An even bigger challenge was that it was impossible to test the system in the usual way. Airplanes and all prior spacecraft were assembled and tested thoroughly on the ground before flight, but ours could not be. The fully assembled station was too big to fit in even the biggest structure in the region, Houston's famous Astrodome. Moreover, even if it had fit, its fragile hardware, designed for the weightlessness of orbit, would bend and sag under the force of gravity, preventing the close alignment and functional testing we needed. Finally, any big facility would be filled with air and pleasant temperatures for the humans who worked there, not the vacuum and extreme temperatures found in space. These factors meant that things might work on the ground but not in space, and vice versa. Unlike any other machine humans had built, the station would consist of pieces that would touch each other for the first time in space, under conditions they had never been under before. Everything had to be perfect, and the potential for unpleasant surprises was enormous.

Many of my new colleagues believed that some of the program's redesigns were due to managers not fully respecting these challenges. Some managers even thought that developing the station would be dirt simple compared to the technological leaps of Apollo and the Space Shuttle program, with some believing that assembly would be a breeze—parts snapping together like a plastic kit.

But other colleagues, especially those with scars from previous programs, were far more circumspect. I made careful note of who was worried about what and why. Later, I spent time with as many of them as I could, exploring their concerns and folding them into my big picture.

Even those who thought designing a space station to be straightforward had to admit that new concerns continued to reveal themselves as we dug into the details. For example, at the beginning of the redesign, the long boom that supported the solar arrays away from the pressurized modules was planned to be assembled by spacewalking astronauts using individual tubes and junctions. Electrical wires and fluid lines would then be routed along this open structure with the aid of deployable utility trays.

However, the analysis showed that this approach was awash in risk due to the hundreds of individual pieces to be handled and connections to be made. Even with the very best design and workmanship, the odds of a major problem were discouragingly high. This is what finally killed this approach—the inability to verify the final fit and function of the critical components before launch. The so-called "sticks and balls" approach to the truss required hundreds of structural joints, electrical connectors, and fluid connections that would be launched without having been verified to work in their final position. If this had been the only way to build the truss, we would have been forced to accept the risk, but it was not.

The redesign team sought to reduce risk by performing as much assembly and testing as possible on the ground before launch. The result was called the pre-integrated truss design, or PIT. Instead of assembling the truss from individual pieces and then stringing utility cables and fluid lines in space, the PIT broke the truss into sections that could fit in the shuttle cargo bay with systems, cables, and fluid lines already installed and pre-tested. The number of connections to be made shrank from hundreds to dozens, and our confidence rose dramatically.

Our work was being closely followed by our program leadership in Reston, Virginia, a suburb of Washington, DC. Both Barry Boswell and I functionally reported to the overall logistics and maintenance manager located there, Jerry Johnson. Jerry was a marine corps reserve colonel, a serious man with a lean, muscular physique. He had been a helicopter pilot early in his career, doing what marine aviators do: taking care of their troops on the ground. He had also flown Phantom jets, and we shared many flying stories. I liked him immediately.

Jerry had gone from the jungles of Vietnam to the world of NASA, where he had worked alongside Barry

Ship and Shoot

One of the management approaches considered to save time and money was to launch hardware directly into space without integrated testing at the launch site. This was in stark contrast to the traditional approach, which involved performing extensive testing on the ground before each component was launched. Budget and schedule pressures gave the shortcut strong appeal, and the testing program remained one of the most contentious aspects of the program.

Boswell on the early space shuttle missions. They had been part of the team that had fought long and hard to give the shuttle crews some in-flight repair capabilities to fix problems that might arise while in space. These included solutions to critical but unlikely problems, such as closing the cargo bay doors if they jammed and limited repairing of electrical systems in the crew compartment. They were determined to bring that experience to the new program, where maintenance was not a remote possibility but a core requirement.

The redesign gave me the opportunity to spend time with Jerry and learn what he needed from MOD. Although his job in the program office revolved around budgets and politics, I could tell he would rather be at JSC doing the nuts and bolts work. We got along well, and I gained a lot from his experience. We soon started talking on the phone on a daily basis.

The redesign introduced me to other people who could make our work easier or harder. One was an up-and-coming engineer named Bill Gerstenmaier. Tall, slim, and taciturn, "Gerst" had begun his career at NASA's Lewis Research Center in Cleveland, Ohio, before joining JSC to work on the shuttle. He was leading the assembly operations planning team and gave daily feedback on the operational impact of design decisions. Although Bill and I would work together closely later, at this point, he viewed maintenance as a minor consideration that did not warrant his attention. Like most of my colleagues, he focused on the specific task he was given and ignored the rest.

Another manager skeptical of maintenance requirements was retired astronaut Bob Overmyer. Bob had flown fighters for the marines before being selected as an astronaut for the military's Manned Orbiting Laboratory program in 1966. When that program was canceled, he became a NASA astronaut—too late for Apollo but finally making two shuttle flights before retiring. He was a senior manager at McDonnell Douglas, the contractor responsible for the truss and other major parts of the station system. We soon became good friends, swapping stories and flying airplanes in our off-hours.

Despite that friendship, we had very different views on the station design and approach to maintenance. Bob was among those who believed our hardware would be super reliable and that maintenance could be deferred until after assembly was complete.

Such different perspectives could lead to conflicts between smart, confident people. While I'd worked at NASA Headquarters, I had been in Space Shuttle program meetings during which managers had stood and shouted at each other with red faces and bulging veins. I had yet to see this in space station meetings, but the potential grew as the pressure to complete the design increased. Despite disagreements, people generally sought to win arguments on their merit rather than emotion.

The pace was relentless, with changes reported daily and critical feedback flowing in from affected parties. The process was not for the faint of heart; many of the assessments were brutal. It reminded me of my time as a fighter pilot, when everyone was a peer, feedback was direct, and participants needed a thick skin.

Throughout the grueling weeks of the redesign, people threw themselves into their work with amazing energy and commitment, taking whatever criticism came as honest feedback. I felt honored to be part of such a dedicated and professional team.

Around this time, I received an invitation to meet with Randy Stone, then the head of the Flight Director's Office. Randy had heard about this newcomer and decided to size me up for himself. After a few questions about my background and goals for the future, he quickly extended a warm welcome and made it clear that he would be delighted to help me learn the MOD way. Such acts of kindness and professionalism were the norm for this extraordinary gentleman.

A few weeks later, Freedom management asked for the MOD position on how the redesign was affecting maintenance. Although that was clearly my job, my latent insecurity led me to hope that my boss, Dennis Webb, would take the first shot. Before I had the opportunity to propose that option, Gene Kranz made it clear that it was time for me to step up and get into the fight.

My reticence was soon overtaken by necessity. My program office superior, Jerry Johnson, was recalled to active duty with the marines in support of Desert Storm, the war to free Kuwait. Although he would not be going overseas, he would be on active duty and unavailable to support the redesign activities.

Jerry called to give me the news. "Jim, you need to take over as the acting maintenance manager for the program. You can stay in Houston, but you will be in charge of the whole team. You will also need to come up here once a week for major management meetings. I'll put you on distribution for everything you need. Don't worry, you'll do fine."

This threw me from the frying pan into the fire before I felt comfortable with even my own part of the work. The next day, I flew to Reston to meet my new bosses and learn their roles and needs. That short visit made it clear that internal NASA political considerations were hurting the design, so I carefully set out to make adjustments in my area of responsibility. As the new guy, I felt conspicuous, but I had no choice. By keeping my changes small but substantive, they slipped through without too much opposition. What I could not judge was how my work was being received by my colleagues and bosses. A few weeks later, I got feedback.

Every morning at eight o'clock, Gene Kranz had a telecon tag-up with his division chiefs spread all around the complex. He would lead off with news from senior management, then ask each office for a summary of what was going on. I normally listened in from Dennis Webb's office. One morning, Gene mentioned that he had attended a meeting during which I had delivered my early assessment of the logistics and maintenance impacts of the new design. Speaking in his crisp monotone, Gene reported, "Jim Van Laak did an absolutely superb job laying out our needs and concerns. He has the full confidence and attention of the program team."

I was thunderstruck. Dennis smiled broadly from across the table, showing his delight with every victory earned by his team. Gene's simple statement left me both startled and thrilled beyond words. Until then, I had been terrified that I might let down the great Gene Kranz, and I frankly dreaded getting my first real feedback from him. Now, I floated down the hall to my office.

I wondered how long the honeymoon would last.

5. EARNING MY SPURS

The words of praise from Kranz marked the beginning of a new phase of my career. The clear support from my management gave me the warrant to speak out on important issues, though I vowed to move carefully and slowly. I was still the new guy, and it would be easy to get myself into trouble.

When Jerry Johnson returned to the program office after Desert Storm ended in the spring of 1991, he stunned me by asking me to continue the maintenance manager job. Because the leadership roles were supposed to be located in Reston, my role was described as "acting," but it soon became clear that I was effectively the level 2 manager.

"People like what you are doing, and I do, too. I've been asked to take on more of an integrating role with the utilization and operations folks and let you focus on maintenance and logistics."

This sounded almost too good to be true. I had gone from being the greenest member of the team to the leader in less than ninety days, and although I would never have sought the role, it allowed me to keep my initiatives rolling. This trust did wonders for my self-confidence, but working so close to my limits kept my ego in check. It helped me see my role as helping a team of high performers find the best answers, not dictating to them.

Embracing this leadership role forced me to address the next questions about how maintenance fit into the larger station program. With every model showing more maintenance demand than could be accommodated during assembly, I wondered how much maintenance could be deferred without putting the station at risk. Was the station design adequate to survive the kinds of problems that could realistically be expected?

My early inquiries into this subject were disappointing. NASA always designed its spacecraft with backup systems to fill in when things failed, and the Freedom Station was no exception. But there were many different ways these backups could be

> ### Levels of Management
> NASA designated the various levels of the program management structure based on their level of responsibility. Level 1 was Headquarters, with responsibility for liaison to the White House, Congress, and other agencies. Level 2 was the integrating program office with all of its functions. Level 3 was made up of the various NASA center organizations and their resources.

included, none of which were intended to protect for months or years after a failure. What might work for a ten-day mission was out of place for a thirty-year spacecraft.

The formal requirement for backup systems only demanded that the station continue to function in the presence of a single failure by a single component. This was considered acceptable because the station was being designed to be maintained, and failed hardware would be replaced as soon as possible. However, because the station had so many operating pieces and a long, slow supply chain, some repairs would take months. This meant that the problem was not just surviving a single failure but surviving the set of failures that could accumulate while waiting for repairs. The result was a practical requirement to be two-failure tolerant so that even after the first failure, there was a backup while awaiting maintenance. A realistic failure rate over so many components could result in enough broken hardware that we could not maintain the critical systems, meaning we could lose the station. I needed to answer the question of whether this aspect of the design was acceptable.

Such questions were the responsibility of the systems engineering group in Reston, but they felt they had higher priorities. When I asked the question, they were dismissive.

"We have to complete the basic design before we look at that."

"But this is critical for surviving assembly," I said. "The early stages are the most dangerous because there are few backups and little margin. How can you assess the design without knowing if it can survive that early time?"

But they insisted they had more urgent, though arguably less important, issues. They were struggling to meet goals that were easy to measure, like weight and cost, with no time to deal with more abstract issues like reliability. After all, NASA was paying a lot of money for these systems, so they had better be reliable!

That answer did not work for me. My experience with aircraft had shown that problems occur with every system, no matter how expensive and perfectly designed. It was our job to be ready for them. Since the Reston group could not help, I decided to create my own approach, working through my colleagues at JSC.

Larry Bourgeois was the senior MOD leader responsible for all operations work on Freedom. Tall, slim, and laconic, Larry had recently left the Flight Director's Office for a management position. I was sure that he would appreciate my concern and have suggestions on how to proceed.

"Not our problem," he said after hearing me out. "Let the program design the station, and we will figure out how to operate it."

I was stunned.

"But what if they give us a design that we can't operate safely? They are counting on MOD's feedback to guide the design, and it is my job to ensure we can keep it maintained. We have a responsibility to the program."

"We will figure it out when the time comes," he said.

I tried a different tack.

"Okay, well, my work would benefit from some operational guidance. It would really help if I had some basic flight rules to work from. On the space shuttle, you have rules that say you have to cut the mission short when certain failures happen. How about some guidance that says we need to be able to restore critical redundancy within a certain period of time, like thirty days? That would give me something to shoot at."

Larry smiled.

"Jim, just let it go. Trust the designers to do their jobs, and then we will do ours!"

> **Attitude Control**
> Unlike objects on the ground that will generally stay put unless disturbed, objects in space need constant attention if they are to stay in one orientation. This function is called attitude control and is made up of many subsystems to detect the attitude, exert forces to maneuver, and stop the motions when the desired attitude is attained, all of which require power, thermal control, and communications.

Having hit another dead end, I started thinking about how to better describe the problem. I knew that the designers were doing good work, but there were so many pieces that the probability of a serious failure was significant. Even though the finished station appeared robust, during its early stages, it was very fragile because so few backup systems were in place.

It was clear that not everything was equally important to the safety of the station. Experiments could fail, and the station would be okay. Even losing some of the life support system might be tolerated if the crew could return to Earth until it could be restored. As long as the station was not vacant for too long, it should not be critical.

But one function was so critical that it had to be maintained at all costs—attitude control. It was imperative that the station be stable so we could dock the space shuttle and fix whatever had broken. The reliability of the attitude control function became my number one priority.

The attitude control design for the completed station was excellent, with extensive redundancy and extra capacity, but during the early stage of assembly, it was extremely fragile. The plan for getting through the assembly period was simply to go fast. The designers thought they could depend on the hardware to work almost perfectly for what they hoped would be a short time.

That was the crux of my worry. If it were possible to assemble the station over a weekend, you could reasonably expect to get through it without too many failures. If it was going to take a decade, you would make exhaustive preparations in anticipation of some serious problems. The Freedom Program was somewhere in that enormous gray area between those two extremes. The program leaders assumed that the complete assembly would take about three years and the vulnerable period of minimum redundancy much less. But because of the complexity of the task and its dependence on a constant stream of shuttle flights, nobody had the slightest idea of how long either one would take.

Although there were many things that could go wrong, my biggest worry was the potential for a problem with the space shuttle, which would cause a long delay. The *Challenger* accident had grounded the shuttle for nearly three years, and many smaller problems had caused delays lasting a month or more. Any serious safety issue with the space shuttle forced NASA to go slowly because it did not have a crew escape system. In a real sense, NASA put the future of the agency on the line with every flight by betting that the vehicle and its operations would be absolutely perfect. Achieving such perfection meant that any significant problem needed to be fixed before launch, and that could take time.

There was another equally serious challenge. Because the orbiting space station would travel over five miles every second, the shuttle had to be launched within about a five-minute window in order to catch up to it in a reasonable amount of time. We did not know if we could consistently launch in that small window of time. If the shuttle could not consistently get off in that five-minute window, the assembly timeline could be extended by several years. Although my formal job was simply to maintain the station, my sense of responsibility made it my job to make sure we could survive a greatly extended assembly phase if required.

The first step was to find any weak elements and get them fixed. Ironically, the most critical function, attitude control, was among the least robust during the early stages of assembly. Seeking a solution, I asked the reliability group at JSC to analyze it and suggest opportunities to improve. They were already reviewing the design, but this gave their work new urgency.

Next, I needed to figure out how to respond quickly to failures that could occur. With no planned maintenance on the assembly flights, the best we could hope for was to have spares ready to load onto the next shuttle with last-minute training for the crew. We needed to be ready to respond if or when a problem arose.

Despite my best efforts, station management was not convinced, and they refused to allocate any resources for maintenance on the assembly flights. They thought it was a waste to reserve resources for something that we could not prove would happen. Instead, they would count on NASA's ability to respond to a crisis if required. Although I knew NASA was good at that, I was concerned that the approach would guarantee a string of crises and add a major risk that we could easily avoid.

By now, I had been at the job for about six months and was just beginning to get a handle on the problem. Then, in the early summer of 1991, I was informed that the General Accounting Office (GAO)—now the Government Accountability Office—would be performing an audit of Space Station Freedom's maintenance and logistics program. The audit was prompted by the findings of the Fisher-Price Study completed the previous year. To me, this sounded like it could be a major burden for my team, but Jerry Johnson told me not to worry.

"They'll make a few visits, ask a few questions, and that will be it," he assured me. Although I had no experience with this group, I suspected something longer and more intrusive.

The audit team consisted of five men and women from GAO offices across the country. They were specialists in a wide range of general program functions, such as budgets and schedule analysis. All were extremely bright and dedicated, but I was distressed to find there were no technical people on the team. Since ours was a very technical issue, I wondered how they would judge the adequacy of our work.

The audit team briefed NASA management on the rules governing the audit, beginning with NASA Headquarters and Freedom management in Reston. They repeated the same briefing to the team at JSC, with me as the senior Freedom manager present. The auditors were very smooth and professional, outlining the goals of the audit and emphasizing their authority to do it.

Following the meeting, the auditors and I adjourned to an office in Building 45 set aside for their use. They were friendly, but being parties involved in an audit is hardly the basis for a pleasant relationship. Adding to the tension, the setting they had created in their office was intimidating, with five chairs arranged in an arc around a single chair at its focus—mine.

They were well prepared, and each carried a copy of the two-volume Fisher-Price report with yellow stickies bursting from nearly every page. They had obviously spent weeks studying the program, the station design, and especially the report. Since each sticky represented a question, many requiring long discussions of technical and operational issues, it was obvious that this was going to be a long task, far beyond Jerry's estimate of a few days' work. I asked the team what they anticipated for the schedule and what support they needed from the program.

"We expect to take about six months. We will be making multiple visits here and to other centers to interview those doing the work. We will also survey relevant experts in other agencies and programs. We will need you to answer specific questions, and we will ask for data to substantiate your positions on key issues. In short, you will be seeing a lot of us."

This was going to impact our work even more than I had feared. Answering their questions would require time and energy from already-overworked engineers trying to stay on schedule. Moreover, with no technical people on the GAO team, I was concerned that they would not understand our explanations, drawing conclusions that were inappropriate.

I resolved to do what I could to minimize the impact on the working engineers, but many of the issues the auditors wanted to hear about would require expertise I did not have, and they would want to hear it directly from the experts to prevent my shading the truth. To minimize disruptions to our fully occupied team, I asked that they come to me first whenever possible. I promised to be completely frank and open, providing data and authoritative sources for our positions.

To demonstrate my good faith, I offered my frank assessment of what their product was likely to be, including our program's weaknesses.

"The Fisher-Price team did a very thorough job, but I'm confident that you'll come to see that their report was extremely conservative. You'll also see that the program takes the report very seriously and that Jerry Johnson was hard at work on solutions before the report was even finished. We are continuing and expanding on his work; in the end, you'll see that there are two remaining risks," I said.

"And what might they be?" asked the lead auditor.

"The first risk comes from the fact that reliability data is statistical and can show large variations in individual units. It is likely that some hardware will be better and others worse than predicted, and we will not know which is which until they are in service. That is the physics of the problem, and we have to deal with it by planning as realistically as possible."

"And the other?" she asked.

"Right now, we have very tight manifesting constraints on the early assembly flights. There's little margin for carrying spares, especially when we can't tell exactly what may be required. So, the baseline plan doesn't have any resources for maintenance on any of the early flights. We are going to have to respond to failures when they happen, and we are going to have to come up with a way for parts and tools to be added late when the need becomes clear. I am confident you will have findings on those two risks."

After the meeting, I gathered my team together and discussed what I had heard. Some wanted to lay out their full wish list so the audit report could bring attention to the issues we had been struggling with. This group saw the audit as an opportunity to gain leverage for what we wanted to solve our problems.

I knew that would be a disaster. Using the GAO against our NASA management would be both unprofessional and damaging to the program. It would also be wrong because this was our problem to solve through sound technical arguments. Our problems constituted just a few of the hundreds the program was facing, and it was our job to convince program management if and when we deserved priority for our issues. It was paramount to protect the integrity of the engineering process against political interference.

However, the audit had a positive aspect; it turned up the heat on solving our problems. Answers we might have found two or three years down the road had to be found within the next few months. Like so much adversity, it was a grand opportunity as well.

The next few months were a whirlwind of good and bad. We made excellent progress on many points and saw the list of potential issues slowly diminish. In other cases, we struggled to communicate arcane technical points that the auditors could not quite grasp. I was concerned that we might win the battle but lose the war.

Then, in the middle of this contentious work, my mother died in an accident. Although she and I had been very close for most of my life, the move to Houston and the pace of the work had loosened the link. As it happened, my mom had called a few hours before she died to say, "Tomorrow is Saint Patrick's Day, and I want you to remember that only good things happen on Saint Patrick's Day." Although I did not know it during that conversation, she'd clearly had a premonition that it would be her last. As a deeply religious woman, that was her way of telling me that she was at peace with her fate. Shortly after midnight, she was gone, and I was absolutely stunned. The next week was a blur as I flew home to be with my family, but beyond that, I was lost. Dealing with my own grief and that of my family took all the intellectual and emotional energy I could muster.

Fortunately, my team continued without missing a beat. The auditors were also considerate and gave me a couple of weeks to regain my balance before resuming their direct questioning.

The audit concluded in late spring, and I traveled to Washington, DC, to accompany the program director, Dick Kohrs, to the outbrief. The auditors had provided me with a draft of the report, which focused on the two findings I had predicted so that I could correct any minor errors. I had also sent the final draft to Kohrs. We met in his office to prepare anything that might be required to answer the two findings.

"Sir, I have to tell you that you did not have good staff support on this," I said. "For months, I have been passing the word that a letter from you making room for some maintenance planning on the early flights would prevent one of the two findings."

"The problem wasn't my staff," Kohrs said. "I didn't want to sign the letter. But I have it with me today."

Kohrs was a very capable engineer and manager, so I knew he had his reasons, but having the letter six months earlier would have saved me a lot of work. On the other hand, handing it over at that point gave the GAO a victory by having NASA concede a point.

The outbrief itself was low-key and professional. The audit team outlined their assessment with its key findings, exactly as I had proposed at our first meeting. NASA addressed the manifesting issue with the letter from Kohrs, and the statistical uncertainty was acknowledged as inescapable. We outlined our plans to manage the risk of the uncertainty, and they acknowledged that it was all that could be done. I was very happy with the outcome.

As we were leaving GAO headquarters, two of the auditors stopped me.

The lead auditor spoke first. "We just wanted you to know that we thoroughly enjoyed this audit and learned a great deal. Thank you for being so patient with us while we came up to speed. You have a wonderful ability to explain difficult topics in plain English."

"Which is how we came up with your nickname," said the other.

"I have a nickname?" I was surprised.

"Roget," they said, "as in the thesaurus. You found an analog for everything, and it really helped a lot!"

Two days later, I was in a meeting with Hank Hartsfield, a veteran astronaut who was leading an important technical review for the Freedom program. As we were leaving the meeting, he pulled me aside to tell me about a call he had gotten from Aaron Cohen, the JSC director and acting deputy administrator of NASA.

"Aaron wanted me to tell you that he is absolutely thrilled with the GAO report. He said it is the finest audit report NASA has ever received! You have hit a home run!"

6. THE MOD WAY

The success of the audit was an enormous relief, and I quickly shared the good news with my team. Although many people had supported the effort, the brain trust was a small team of contractors led by Julie Bassler, a quiet, friendly, and extremely sharp young engineer at McDonnell Douglas. The volumes of data she and her team produced showed the GAO that we were on top of the problem.

As news of the good report made its way around the center, the attention it received brought an unexpected complication. Some seniors in MOD began to worry that I was speaking for the operations community without having enough experience. Larry Bourgeois, who had dismissed my concerns about the design, was particularly anxious. He proposed that I spend six months on detail to the Flight Director's Office, and Gene Kranz agreed.

The public sees the flight director as the individual in charge of the flight control room. He or she has complete authority to make routine decisions that fall within management-approved guidelines called flight rules. However, because things do not always go according to plan, the flight director also has emergency authority to take any safety-critical action when there is no time to consult with higher management. It is a role with extraordinary responsibility and authority.

Although I understood the role of the flight director, Larry wanted me to deeply understand not just what they did but how they did their work. At first, I thought this was overkill and was not excited about the decision to send me over there. Dennis Webb quickly corrected my attitude, making clear that it was an extraordinary honor and that I should behave accordingly. Properly chastised, I resolved to do my best and quietly hoped that it was an indicator of good things to come. At a minimum, I knew that learning the flight director's business would give me a better understanding of mission operations for the shuttle, which would be critical during assembly flights, as well as insight into how things might need to change for Freedom.

My friend Randy Stone had been replaced as leader of the Flight Director's Office by Lee Briscoe. Like all flight controllers, Lee was extremely intelligent, though not as welcoming to outsiders as Randy had been. However, since my detail came with Kranz's sponsorship, I was quickly accepted into the group and immersed in their planning and training process.

Each shuttle flight was assigned one flight director as lead, and it was their job to take the flight from "cradle to grave." This included considering any contingencies that could arise, such as hardware failures, bad weather, radiation events, human error, and many more, and making preparations and allowances to deal with them should they occur. The process grew out of aviation flight test operations pioneered by NASA's predecessor, the National Advisory Committee for Aeronautics (NACA), and matured over many space missions.

This exhaustive planning was mirrored by the engineering support team. The two groups produced precise estimates of fuel usage, power margins, operating temperatures, and many other detailed parameters to ensure that the flight would work out with the desired safety margins. Telemetry from the vehicle would show how well these predictions were met, allowing real-time adjustments if needed. After the mission, the engineers would look for anomalies needing further study. This was the work required to get the reliability needed to safely fly people in space.

The JSC teams were also supported by technical teams across NASA's other centers, the contractors who built the vehicle, and the people at the Kennedy Space Center who prepared it for launch. Together, they formed an enormous network of experts capable of analyzing all aspects of the vehicle and its performance. A few years later, the Mission Evaluation Room (MER) in the basement of Mission Control would become as familiar to me as the living room in my house.

While I had long appreciated the talent of the operations and engineering team, this was my first opportunity to be a part of their work. Embarrassed by my initial reluctance to take the detail, I quickly recognized the enormous value of the experience I was getting. My days were filled, adding reviews of every aspect of upcoming shuttle flights as plans were scripted, trained, and tested, to my work on Freedom maintenance. The detail and precision of the work were extraordinary because MOD's people are perfectionists, and nobody wanted to let down the team or be caught unprepared.

Although these were many of the smartest people in the agency, it soon became clear that being smart was not enough. The complexity of the problems meant that they often could not be solved in real time. Instead, the team spent many hours ahead of the flights identifying potential problems, after which detailed analyses were done to carefully weigh the risks and develop the best plans of attack. These became flight rules, and they were the bible for mission operations.

Flight rules were created through a process called "flight techniques" that ground through these potential problems in excruciating detail well before they might be encountered. The process gave everyone time to consider possibilities with a cool head, conduct analyses, and discuss options. It was also the ideal forum for learning about how the systems worked, although the extreme attention paid to the minutest details sometimes made staying alert a test of professionalism.

Since I was just an observer with no formal responsibility, I sometimes struggled to stay engaged. To overcome this challenge, I developed a habit of imagining myself as the decision-maker. It required me to mentally analyze every detail, challenge every ground rule and assumption, and strive to understand every aspect of the problem and its solution. This role-playing greatly aided my learning. It was also enormous fun.

One day, Kranz asked me to come by his office to talk about my detail and what he hoped I would learn from it. It was his last meeting of the day, and after covering my work, Gene went on to share some personal experiences from his career, including one that helped form the MOD culture. The story is almost legendary, and hearing it directly from him made an indelible impression on me.

Gene wanted everyone in the organization to be a part of the MOD team from the core of their soul, and he knew no better way than to tell the story of how the Apollo 1 pad fire that had killed Gus Grissom, Roger Chaffee, and Ed White had changed him and the operations culture. The mishap had been a watershed event for NASA, and it was critical to the success of all that followed.

The accident had happened during a test at the Kennedy Space Center and had not directly involved MOD. Most flight controllers had been at their consoles in the Mission Control Center to monitor the events for any nuggets of information they could gain, but they had not been directly participating in the test. They had heard the tragedy unfold in real-time, utterly helpless to do anything but listen in horror.

The Apollo program, which had been going a mile a minute, had suddenly frozen. All activity toward flight had been stopped. Morale had cratered, and Gene had known he needed to act quickly to learn from the tragedy, calm the inevitable fears, and restore the team's confidence. He assembled the flight controllers in the auditorium next to the Mission Control Center.

As he had with hundreds of others before me, Gene shared the message he had given his team that day. It went something like this:

"Folks, we failed. We failed the crew, and we failed the American people. We knew that there was a risk of fire in the cabin, but we said nothing. We assumed that others had the risk under control, and we were wrong. That will not happen again. From this day forward, the motto of the mission operations team is 'Tough and Competent.' 'Tough' means we will accept any consequences that come from doing the right thing, and 'competent' means we will know all that we can possibly know about our systems and our disciplines."

Everyone who had been present that fateful day had understood the meaning of his words and the emotion behind them. A few of his colleagues later said they thought he was being melodramatic, but the result was profound. The MOD culture that I experienced was the product of that attitude, and even today, more than thirty years after he shared it with me, recalling that conversation gives me chills. Gene turned generations of cocky young engineers into a tight and powerful team. His tools were leadership, integrity, and professionalism, and his tale had the desired effect on me, galvanizing my resolve to be as professional as I could possibly be. It also gave me a new and palpable appreciation of MOD's culture.

That insight helped me both understand my fight director colleagues and measure them against the model Gene had described. At the time, the office held about a dozen flight directors, each with a strong and distinct personality. All of them were very smart, and some were brilliant, but most appeared arrogant and conceited, especially when they were competing among themselves in the office. Once I got to know the individuals and their work, I realized that some of that arrogance came from their efforts to live up to the role. For the best of them, it was their way of using their forceful personality to pursue perfection.

Because my focus was station operations, I was assigned to follow Bob Castle, a mild-mannered gentleman who specialized in the on-orbit phase of the mission, which was most similar to station operations. He made me his shadow, taking me along when he met with the crew and controllers, sharing his console in the flight control room during missions and simulations, and immersing me in the preparations for his upcoming missions. He was very gracious, sharing what he was doing and why.

In addition to the technical knowledge I gained, the experience also highlighted the caste system in the office, with the launch and entry flight directors on top and the others below. As a fighter pilot, I knew that status-seeking was natural in high-performance organizations, but it was not always helpful and was often unpleasant to watch. Sometimes, it could even distort personalities and affect judgment. On rare occasions, this caused me to question the quality of some people's decisions.

The actions of my flight director colleagues reminded me of an example that had happened while I'd been working for Michael Greenfield at headquarters. A space shuttle landing at Edwards Air Force Base in California had been given a bad setup for the final approach, well short of the normal path to the runway. Although the shuttle had landed safely, the error had been real and could have had a tragic outcome. When I'd been sent to talk to the flight director who had overseen the entry, he had refused to even look at me. Throughout the hour we'd spent together, he had never looked at me once, preferring to stare out his window. To this day, it remains one of the most egregious demonstrations of professional arrogance I have ever seen.

That same individual was still a flight director during my detail, and we spent some time sitting side by side during simulations. He never hinted that he remembered me, and I never brought up the incident, but it caused me to carefully scrutinize his judgment for as long as I knew him. Eventually, my concern was justified by his role at the time of the shuttle *Columbia* accident.

I got another demonstration of such ego, albeit less extreme, when another of the ascent flight directors came by and stuck his head into my cubicle. "Jim, want to see what a *real* flight director does for a living? I am going over to a shuttle meeting to talk about lake bed landings. You are welcome to come."

As we walked across the lush JSC campus, he filled me in on the issue.

"We have contingency plans that tell us to land on the lake bed rather than the runway at Edwards if we have any concerns about the entry or vehicle. The lake bed gives us a much larger landing area to tolerate human and system errors. The only downside is that the natural surface can be very soft after a rain. This meeting is to talk about some testing that could help us understand the risks of landing there."

We joined a group of senior space shuttle managers in a fifth-floor conference room. The staff was preparing a recommendation for the program manager and wanted to review every aspect of the issue. It was a comprehensive briefing to provide their best recommendation to the boss.

After the engineers made their presentation, the flight director boldly spoke up. "There's no point in doing any of this," he said, "because I wouldn't believe the results anyway." His tone was declarative, as if his judgment was absolute. What struck me was not his technical opinion, which was reasonable if not convincing, but the manner he used to express it. This was not the behavior we were taught to expect in high-stakes meetings, especially in our post-*Challenger* NASA.

The group ignored his outburst. They thanked him for his opinion but made it clear that they were going to go forward with a recommendation to proceed. The program manager agreed and, over time, added additional investigations to better understand the pros and cons of using the lake bed. His decision struck me as exactly correct, and the flight director's arrogant dismissal struck me as dead wrong.

My experiences in the Flight Director's Office—and the intimate interactions with crewmembers they required—changed the way I looked at these key participants. Their intellectual capacity was stunning, but they were as human as everyone else, which could sometimes hurt their decision-making.

Flight directors and astronauts were also prone to other human frailty. Because of their real or apparent power, others wanted to be in their good graces, often leading to exuberant displays of admiration from their minions. Such fawning attention encouraged some to become so enamored with their own brilliance that they could overestimate their expertise. This could be very embarrassing, though it rarely resulted in serious problems.

For all their power and celebrity, flight directors and astronauts were actually pretty low in the management hierarchy. Responsibility for the overall success of the program rested with more senior managers who, while appreciating the input of these experts, rarely deferred to them.

In many ways, I found the Flight Director's Office like a fighter squadron. Its members were sharp, arrogant, and quick to step into a leadership void. But there was a critical difference: Fighter pilots put their lives on the line and often scare themselves, an experience that moderates most egos. Flight directors work in the safety and comfort of the Mission Control Center, where they neither feel fear nor experience danger. Too many were oblivious to the times when their judgment was weak; they felt vindicated as long as no actual mishap occurred. The deep introspection that followed my plane accident as a teenager had taught me the difference between getting away with something and being right. Most of the flight directors had enjoyed no such introspection.

Although I was just there to observe, I soon felt that I fit in and could learn to do the job very well. The work was complex, but the support from MOD and other experts was superb. There were even rumors that I was being considered for the next class of flight directors, an extraordinary honor for someone who had not been a flight controller first.

But fate took me in another direction. Headquarters aborted my detail after a few months and called me to Reston to help Freedom through yet another technical and budgetary challenge. That effort was successful in achieving modest cost savings, but it demonstrated how fragile the support for the program was. A few months later, events would take the program in a direction that would change NASA's future forever.

7. ONE MORE REDESIGN

The election of 1992 brought a new president to the White House. Bill Clinton had a keen focus on the economy, including a promise to reduce the budget deficit and control government spending. One of the programs that caught his critical eye was Space Station Freedom.

On a lovely morning in early 1993, I was meeting in my office in JSC's Building 4 North with one of my contractors, Frank Sager. We were discussing tools to make EVA maintenance easier and faster. The NASA administrator, Dan Goldin, was going to hold an important press conference, so at the appointed hour, we adjourned to watch him on a TV in a common area.

Mr. Goldin was very serious in his address. He said there was good news and bad news about the future of the agency. He described his recent visit to the White House during which he'd been told that Space Station Freedom was not supported by this administration. The president considered its cost excessive and its value insufficient in such times of tight budgets. The president intended to cancel the program.

The room fell as still as a crypt. Despite the many problems that Freedom had experienced, we were making progress at last, starting to build real hardware and developing operating plans to make the station a reality. This was no time to pull the plug!

But Goldin, an energetic and tenacious leader, would not give up easily. At his insistence, NASA would be given the opportunity to reduce the cost of the program and get it back in balance with the administration's priorities. We had ninety days to produce three options to reduce its cost and increase its value.

As the speech ended, people wandered away in a daze. Frank and I retreated in shocked silence to my office, struggling to understand what this meant for our future. As we pondered our fate, Frank's voice started to waver, revealing more emotion than he wanted to show. As a trusted colleague, he honored me by sharing his personal story—one of many that show why the space program is so precious.

Frank was in his mid-thirties, a very competent engineer, an excellent coworker, and always lighthearted and fun to be with. He positively glowed with delight to be working on the space station for NASA. Used to his constant optimism, I could see that he had suffered a crushing blow to his heart. His familiar voice soon became distant, strained, and filled with heaving sighs. Tears welled as he told me how he had come to work for NASA.

The company he worked for was called Oceaneering Space Systems, a division of a larger company that had its roots in the offshore oil drilling industry. The space systems division was the high-tech wing of a company that was largely blue collar, doing the dirty and dangerous work of building and maintaining industrial oil platforms offshore. In fact, Frank had begun his professional life as a diver working on oil platforms off the coast of California. His work was extremely hazardous, and he saw his share of close calls. As a husband and father, he was glad when the opportunity had come to move up from being a diver to being an operator of the underwater robots also used in the work.

But while driving a robot was less hazardous, it shared many of the same burdens as being a diver. He spent long periods away from home, living on austere oil rigs, working long shifts, and often resorting to drinking to break the boredom of the off-hours. It bothered him greatly to see his life slipping into an aimless, potentially destructive cycle. He thought he could do better, and he ardently wanted to.

One day, he saw a NASA press release showing robots being used to build and maintain NASA's proposed space station. He desperately wanted to be part of the effort, but he knew that an oil field robot operator with no college degree stood no chance. NASA needed intelligent, ambitious people with education and discipline, and competition for those jobs would be fierce. If he was going to join this high-tech program, he would have to go back to school and complete a degree in engineering—a difficult task while supporting his growing family.

But Frank made it happen. Like so many others, in NASA's courageous vision of exploration, he found himself, and with that discovery, he set out to do something meaningful with his life—to serve his country and make a better future for his family. He worked and went to school until he earned his degree in mechanical engineering, whereupon his talent and dedication were rewarded with a new job working on the space station. Because the biggest demand for robotics was maintenance, he found his way to my team. The time we spent together was always fun and informative, and I considered him one of my most valuable assets.

"Now it's all falling apart," he said as tears welled in his eyes. "All my work and planning is just slipping away, and there is nothing I can do about it!"

I struggled to find something positive to say.

"Frank, I have no idea what is coming, but I do know that you have done spectacular work for us—for me. Don't give up yet. For some crazy reason, I still have hope that we can recover from this. Either way, I have enormous regard for your abilities and work ethic."

His wan smile reflected his appreciation.

Within a few days, the new work plan was announced. There would be three teams, each with a different design approach. The Marshall Space Flight Center in Huntsville, Alabama, would lead the first team to develop Option A, the Langley Research Center in Hampton, Virginia, would lead the second team to develop Option B, and JSC would lead the third team to propose Option C. Although I was repelled by the thought of another redesign, I felt a responsibility to make it work and intended to volunteer to help anyone who would have me. As the expert on maintenance, I was soon asked to support all three teams, but I found my home and made my greatest contributions with the team from JSC.

The JSC redesign team would be led by John Aaron and the JSC director of engineering, Henry Pohl. I asked Dennis Webb for permission to support their organizational meeting. He agreed readily, and I headed off to find the Option C team in Building 15, a complex of offices and laboratories on the west side of JSC.

I arrived to find the conference room filling with a mix of familiar and new faces drawn to this moment. Most of the people I recognized were from engineering, with a scattering from other organizations. I was the only one I recognized as currently working for mission operations.

Henry and John worked together with the gentle familiarity of colleagues who had traversed many years and challenges together. They were both stars of Apollo and brought the perspective of thirty years of space history. John's presence was particularly noteworthy: For political reasons, he had recently been replaced as manager of the Space Station Freedom Work Package 2 organization at JSC. Everyone would have understood if he had said, "To hell with the station," but instead, he was here to help once again.

A quiet and intensely thoughtful man, John was a legend at JSC. He had been a part of Gene Kranz's famous White Team, the flight control team that landed the Eagle on the moon during Apollo 11. Kranz was known to call John Aaron "the smartest man in mission control." Others knew him as the "hero of Apollo 12" for his incredible response when the Saturn V had been struck by lightning shortly after liftoff. On that occasion, the crew was left with only emergency systems operating in the capsule while the rocket thundered its way toward orbit. John had to disregard the chaotic data on his screen and review the system in his head to decide what to do, all within the too-few minutes allowed before commanding an abort that would pull the capsule away from the still-climbing rocket.

Incredibly, John not only figured out why his data was in chaos but also told the crew how to find and operate a switch they had rarely been trained to use, all under enormous time pressure. John's action restored the instrumentation so the crew

> ### Aborts
> The flight control team constantly monitors the health of our spacecraft. Their close attention to its performance could mean the difference between solving a problem, losing a mission, or the death of a crew. During critical times, such as launch and ascent, they must provide the flight director and crew with the right action to take. During ascent, there is little time for responding to failures, after which the mission will be aborted.

and flight controllers could see what was going on. Once the rocket made it to orbit, John helped check out the spacecraft before they had headed off for a successful landing on the moon. A few months later, he would play a critical role in bringing Apollo 13 home from its aborted mission. It was absolutely thrilling to work with such a legend.

Henry Pohl was also a pillar of the JSC community, although a newcomer would never have suspected him of the role. Quiet, unassuming, and piercingly insightful, Henry knew more about engineering than many universities. He had begun his career with the army at the Redstone Arsenal in Huntsville, Alabama, as an enlisted soldier. Although he was a college graduate when drafted, he was destined for grunt work until he struck up a conversation with the civilian engineers at the base. They soon recognized his engineering skills and put him to work in a laboratory rather than hauling artillery around the battlefield. When his enlistment was up, he became a civilian on the engineering staff, working his way up to become the laboratory chief. When NASA was created, he joined the team of rocket engineers from Redstone who became a part of the civilian space program. As a native Texan, he'd found his way to Houston when the Manned Spacecraft Center was formed. The Manned Spacecraft Center later became the Johnson Space Center in honor of President Lyndon Johnson.

Looking around the conference room at the mix of faces, young and old, I was reminded yet again how adversity can bring opportunity. John and Henry quickly and quietly created a team of people eager to work around the clock, framing a proposal for a new, lower-cost station. Like the rest, I counted myself fortunate to be part of the team. My expertise with the issues of maintenance and logistics, although not drivers for the initial design, would be important to ensure the design would pass muster. Everyone was well aware of the concerns that had driven the GAO audit and wanted to prevent them from reappearing.

Each day during the redesign began with a telecon with the leadership team in Crystal City, a complex of office buildings in Virginia across the river from Washington, DC. This telecon was the most difficult part of the day, as the leaders there were constantly reacting to changes coming from NASA Headquarters. Often, the direction would countermand previous instructions; one day, we would be told to focus the station design in favor of materials science research, and the very next, we could be told to shift to favor biological research on animals and human beings. Sometimes, the direction was so specific as to identify individual design features, such as details of the cooling system, apparently for political reasons. Such decisions should have been based on technical feasibility rather than politics and internal competition. This chaos caused an enormous amount of work to be wasted. We wondered about the sanity and competence of the people directing these counterproductive changes.

Each day after the telecon, John and Henry would do their best to make sense of the latest guidance. They looked for what could be reused from previous work before trying to address the new items. I shuddered to think what it was like working in the Crystal City office as such terrible direction flowed out to the field, and I thanked my lucky stars for being spared that agony.

In addition to the redesign, other teams were conducting background research on Freedom. One of them was led by astronauts Jim Wetherbee and Bill Shepherd (Shep) and included my MOD colleagues Jenny Stein and Chirold Epp. They asked to interview me during a four-hour telecon on a Saturday morning. Instead of talking about the Fisher-Price maintenance issues, they asked about the weakness I had found in the attitude control system.

"In my view," I told them, "the program did not put enough emphasis on surviving early assembly. The early avionics needed for attitude control have been simplified to the point of being fragile. There are two long strings of hardware that can only tolerate one failure each. Given the length of time they are needed, I feel the risk of losing the station is excessive."

Shep asked, "What does that mean?"

"According to the analyses done by the JSC reliability group, the probability of surviving the planned time is only about 35 percent. Although that is a very conservative assessment, it makes me extremely uncomfortable because if we run into any problem that extends assembly, like a shuttle launch delay, the problem gets worse very quickly. And don't forget that the consequences of losing the function are catastrophic—losing everything that was in orbit."

"How can that be?" Shep sounded incredulous. "What does the program say to your analysis?"

"The designers say that everything will be tested enough not to fail. I disagree, and I was able to get a few people exploring alternatives to improve reliability. It is a situation screaming for a better approach to managing the risk."

Months later, I discovered that my analysis was used as part of the rationale for the cancellation of Freedom, despite the fact that work was underway on a solution. It is likely that some of the input came from this conversation.

Our ninety days of redesign ended in June 1993. Reports were created and handed off to the review committee that the White House had selected for scoring and decisions headed by the president of the Massachusetts Institute of Technology, Charles Vest. Many of us assumed our summer would be a period of relative quiet and uncertainty as we waited for the outcome. I planned to take some time off to be with my family after the mad crush of the spring.

As the Option C team was wrapping up its report for Washington, John and Henry pulled me aside.

"Jim, we really like the way you work with the team. If our option is selected to go forward, we would like you to run mission operations."

I was floored! "Are you kidding?" I asked in genuine shock. "I have been only a minor player in the work of the team. Surely you can do better!"

But John insisted. "We like the way you think and the way you lead people. You are quiet and levelheaded, building confidence and cooperation in the team. I also saw the way you managed the GAO audit and put those political issues to rest. We are confident you would be an excellent leader for our team."

While thrilled by their confidence, I quickly put the thought out of my head. Within days, it was announced that Option A from Huntsville had been chosen, and in an instant, our JSC team was no more. It would take several years for the new program to be given a formal name, but some started calling it Space Station Alpha.

We were told that the main reason Option A had been chosen was that it was significantly less expensive. The savings came by eliminating the propulsion elements used by the other options and replacing them with an off-the-shelf satellite core. The so-called Bus-1 provided propulsion, power, and early communications in a highly reliable satellite platform that also resolved my concerns about attitude control for the Freedom design. While it made sense at one level, many of us were concerned that the idea was so immature that it would eventually create problems, costing at least as much as it purported to save.

The following week, I was winging my way to Reston to be a part of the last Freedom program review. It would capture the work that had already been done so that some of it could be reused in the new program.

The lush beauty of northern Virginia was always a delightful reprieve from the dusty monotone of the Houston area, and I looked forward to spending time with this amazing group of friends and colleagues. Although we were sorry to be wrapping up our work on Freedom, we were proud of what we had accomplished.

In the midst of our work, I was surprised to get a call from Bill Shepherd, the acting head of the transition team for the new program. He wanted me to come see him the next day at the transition team headquarters in Crystal City.

I met Shep in the same offices where the redesign team had worked through the spring. He and I had known each other casually for a few years. Like me, he flew light airplanes and liked to build things in his well-equipped shop. An astronaut and Navy SEAL, Shep had made a name for himself during his astronaut selection interview. The story has it that when asked by the selection committee what he did best, he had quietly replied, "I do many things well, but what I do best is kill with a knife." Such quiet bravado was common among the male astronauts, and the brass loved him for it.

But this was a different situation. Bravado was neither required nor helpful. Although he had led a small project as a SEAL, he was not a program manager and had no experience with large programs, so I had no idea what to expect.

"I need your help, Jim," Shep said.

"Glad to help any way I can. I'm in town until the end of the day tomorrow."

"No, you don't understand," he said. "I need your help until October. I want you to be part of the transition team."

Once again, I was thunderstruck. The whole redesign experience, particularly the morning telecon led from this very facility, had been so frustrating and unprofessional that the last thing I wanted was to be associated with its progeny. And if Shep, with no experience or credentials, had been entrusted with leading the activity, what did that say about its prospects? It sounded like a fool's choice to tie my future to it.

"Whoa! You don't ask for much!"

He went on. "You have impressed folks with your understanding of risk and how to manage it. If there is one thing this program has enough of, it's risk! We need your help."

His serious demeanor was obvious, but I needed time to think.

"I agree with that, but the next few months are going to be hell, and not just for me. I have hardly seen my family for months. Can I give you my answer tomorrow?"

He nodded. "That will be fine."

That evening, I called Michael Greenfield, the man who had hired me at NASA and one of my oldest friends at the agency. I told him what had happened and expressed my reluctance to become so tightly tied to the new program.

He said directly, "There are two kinds of people in this world: those who come when they're called, and those who don't. This is the time for you to decide which of those you want to be."

His words hit me like a baseball bat, but I instantly knew that he was right.

The next day, I called Shep.

"I will support your team on two conditions," I said.

"Okay," he said.

"First, you are damned right that this is a high-risk program, and if I am going to be part of it, I want your assurance that you will listen to what I have to say. If I am not going to be a valued member of the team, I don't want to be there."

"You have my word," he said quietly but firmly.

"Second, I want the upcoming Fourth of July weekend to be with my family. The spring was tough, and taking on this role will be much worse."

"You've got it!" Bill said with audible relief. "Please try to be back on Tuesday."

Houston seemed strangely distant as my plane circled to land that evening. Looking down at the festive lights of the Six Flags Over Houston amusement park reminded me of the evenings we had shared there when JSC had reserved the park for its employees. The fun and tight-knit community of those evenings seemed so distant as to be from another world, one that I would never see again. I had a deep foreboding that my life would never be the same.

8. ENTER THE RUSSIANS

When I returned to Crystal City after the holiday, my emotions were a stew of embarrassment, anxiety, and grim determination. Because this was a critical time for the agency, with our office at its focus, part of me felt honored to be asked to participate. But another part of me was angry and embarrassed about the mess that had been the redesign and horrified that I might be embroiled in another round of the same hideous process.

It did not take long for my worst fears to be realized. As far as headquarters was concerned, the work of the transition team was literally an extension of the redesign, and they mistreated it the same way. Our charter was to turn the product of the redesign into a plan that could be executed successfully. It was painfully apparent to those of us who had been in the field for the redesign that any successful program would require an effort that was much better focused, with less wasted effort and confusion. Unfortunately, this was not to be, at least not in the beginning.

Despite our misgivings, we threw ourselves into the effort with everything we had, as if our sheer dedication could compensate for poor leadership. Our days began early with the same telecon that had so embarrassed me during the redesign. The only thing that had changed was that I was now close enough to see what was going on. Acting program manager Bill Shepherd, universally known as Shep, would use the telecon to get updates from all the participating centers and share new guidance from agency leadership. To my horror, the guidance he was passing on continued the wild swings in technical direction that we had seen during the redesign, effectively preventing any movement toward closure.

Although I knew it was not his fault, I soon became angry at Shep for continuing the chaos we had endured during the redesign. I knew he felt helpless—and he probably was—but his actions made a successful outcome all but impossible. Any small gains we made one day were undone the next.

Many parties were at fault, but for me, the most vexing were those who claimed to represent the researchers. Rather than collaborating to find common ground, they were in competition, each trying to gain an advantage over the others. We struggled to make the station serve all the communities without choosing favorites, but that meant that nobody would get exactly what they wanted. Some were willing to reduce their requirements for the good of the whole, but many others fought tenaciously to get everything their way. The near impossibility of achieving compromise, even within this group of extraordinarily intelligent people, was both disappointing and an ominous harbinger of the fights to come.

We on the transition team had neither the desire nor the authority to favor one discipline over another, so we struggled to craft a compromise. We proposed buying some facilities over time, spreading out the cost and launch requirements. We

The Power of Venue

This was my first direct experience with the toxicity of working in such a politicized environment. This well-known problem is why companies and agencies conduct their management retreats away from their home base. We learned to take advantage of this by bringing the Russians to the Kennedy Space Center for shuttle launches. The combination of distance from their home environment and the professional demands of a launch helped to put our meetings on a much more productive footing.

could wait to schedule some activities until the station was complete to minimize conflicts. Technology could reduce some technical issues—for example, it could provide special racks to reduce the vibrations that were seen as a threat to materials research, but this added cost and time to the schedule.

Despite our best efforts to find a compromise, too many parties remained inflexible. Gradually, it became clear that a politicized process at headquarters was making normally reasonable people rigid and dogmatic.

The competitive tension became a full-blown crisis at the end of July 1993 when the station design, rather than settling down, began to diverge. Since we were due to present an updated plan to the White House in September, we needed an agreement on a basic station design soon. But instead of narrowing in on a generally accepted answer, the battles for supremacy became more animated. Day by day, the tide went back and forth, with no group gaining the upper hand and none willing to concede its interests. The resulting changes cascaded across the entire design, reopening decisions made during the initial redesign effort and making progress impossible.

Shep was doing his best to lead us, but he could not create the stability we needed. I did my best to be a good soldier and work through the problems one by one, but I struggled to be patient with him. I knew he had no power over the competing interests, but he also lacked the experience to organize this churning froth into a viable system design. Finally, after my gentle suggestions had failed to get his attention, my anger and frustration boiled over.

"Shep, we're spinning our wheels! Half our work is thrown away time after time because the requirements keep changing. There's no way we're going to meet our September deadline!"

"Don't you think I know that?" Shep said angrily. "Headquarters keeps making decisions that we should be making here. We're just puppets in their process. Congress and the researchers are pulling every string they can find to get what they want!"

Shep was clearly as frustrated as I was, but he didn't know how to force the mob to converge. I felt I did.

"And they always will, but part of our job is to force them to get real. We have to! These decisions are not independent. When we add or subtract the alpha joints, the impacts ripple through everything else. The same is true for the thermal control system and how many crew members we can accommodate. There's a structure to these decisions driven by physics, and if we do our job right, we can use the logic it provides to narrow down the options one step at a time. We need to put together a decision tree that shows everyone that we have got to make our choices in a structured, rational way. Otherwise, we're just wasting everybody's time!"

"You keep saying that, and I understand your point, but that takes resources we need for other trades!"

"There can be *no* progress until we do this!" I said forcefully. "Postponing it only costs us more dollars, more time, and more political support. We'll keep spinning our wheels with work that's going nowhere. How many times do we have to go over the same ground and get the same answers?"

The frustration was clear in my voice, and I could see it wasn't helping the situation. I took a deep breath and tried to stay calm.

"Shep, I know everyone's busy, but we've *got* to get some logic into the discussion, or all our work will be in vain. This is not a nice-to-have. Progress is *impossible* until we stop wasting time on dead ends and make the design converge!"

Shep could see that my anger and frustration matched his own, and he could see the logic in what I was saying, but he also did not feel he was in control. He had long since lost his innocence about what it took to run a big program and knew he was in over his head, but it was his job, and he was giving it his all. He also realized that the course he was on wasn't working, and it was time for a change.

"Okay, okay, okay! We've got to do something 'cause this sure as hell isn't working. I'll ask Brenda to come talk to you. Work with her to see what it would take to put something together on this."

Brenda Ward, a bright and energetic woman from JSC with a PhD in experimental physics, had been a part of many studies for NASA. When I explained the goal to her, she saw the logic immediately. She pulled together a small team, and they quickly built a map of design choices that showed how individual decisions led to a design. She and her team did a great job, and her tool, aided by the pressure of our impending deadline to brief the White House, helped the design settle down by the end of August.

Meanwhile, another dimension of our work threatened to overturn everything. During the spring of 1993, we heard rumors that the new program would include the Russians. The prospect evoked reactions among the team ranging from enthusiasm to disgust. My own reaction mixed lingering mistrust from the Cold War with concern that they would introduce a slew of practical problems we were poorly equipped to overcome. Russian space vehicles reflected a very different technical culture that produced different designs and operational approaches. I was not at all sure we could blend these different approaches into one craft.

The proposal to add the Russians was radical, but it didn't come without warning. Soon after the fall of the Soviet Union in late 1991, space advocates had suggested a joint US-Russian activity as an extension of the Apollo-Soyuz Test Project of 1975. In that single flight, NASA had docked an Apollo spacecraft to a Soyuz capsule as a sign of goodwill. Now that the Soviet Union and the Cold War were no more, a new effort called the Shuttle-Mir Program was being planned to fly one NASA astronaut on the Russian Space Station Mir and at least one Russian cosmonaut on NASA's space shuttle. Although progress was slow, work was underway to make it happen.

But while this proposal was widely popular, anything more ambitious was likely to be threatened by the chaos we saw on the evening news. The Russian economy and society seemed to be on the verge of collapse. NASA had been sending people to Russia for the Shuttle-Mir collaboration, and each group came back with new stories of chaos.

I was deeply concerned that even if we found a way to bring our systems together, there might not be a Russian program to join us.

By the beginning of August 1993, we were hard at work assessing the challenge of including our former adversaries. A group of NASA engineers in an adjoining building was meeting with Russian engineers to discuss what could be done. Although I wanted to explore the possibility of a joint program, Shep kept me busy working out program office roles and responsibilities, budgets, schedules, and the myriad of other details required to kick off a multibillion-dollar effort. He also wanted me to find a way to manage all of the risks to the program's success. Risk was one thing we had in abundance.

One afternoon, Shep stuck his head in my office and said, "Dan Goldin is on his way over. He wants to know what the risks are of including the Russians in the program. I told him you were working on that." It was on my to-do list for sure, but now I had twenty minutes to formulate an answer.

NASA Administrator Dan Goldin was a fiery leader who pushed everyone to the limit of their comfort and often beyond. That was no accident, as he wanted NASA to do great things and was frustrated with the many technical and political boundaries of our work. Not one to mince words, he could terrify people who were wrestling with extremely challenging problems. We all dreaded having to brief him on our modest progress, especially since much of the blame lay with his own headquarters staff.

Shep and I met the administrator and the acting deputy administrator, George Abbey, in Shep's office. As was his habit in those days, George sat in the back and made no sound as Shep briefed Dan on our long list of concerns. Shep pointed out that the constantly shifting requirements made it hard to nail anything down and briefly mentioned the decision tree that Brenda was developing to force convergence. Dan was clearly not happy, but he understood the process and accepted that we were doing as well as could be expected. When the subject of risk from Russian participation came up, they turned to me.

"There are many risks to including the Russians," I said, "but two that stand out. The first is whether we can reconcile their technical standards, such as their factors of safety, control architecture, and other engineering details, with our own. Because of the total separation of our economies, their technology and engineering conventions have evolved very differently from ours. They would need to work together flawlessly, not in some half-baked manner."

"What is your assessment of the risk?" Dan asked.

"It appears manageable. My quick review of the publicly available data shows that they have faced the same challenges we have and, in many cases, have found similar solutions. We will certainly have problems to solve, but I suspect that there are good, if not elegant, technical approaches that will work. However, doing so will take time and cost money."

"We will do what we have to do," Dan said. "This is a deal-breaker for the White House, so we have no choice. What is the second risk?"

"The second is whether they will be there when we need them." Dan sat back and nodded slowly as I went on. "With all the political and economic turmoil inside Russia, we must be prepared for the possibility that they will be unable to meet their obligations, even if they want to. That means we have to be extremely thoughtful about the role they play in the program, especially what hardware and services they provide. We have to prepare for the case where they cannot continue for any reason."

"That is the real risk, isn't it?" Dan spoke quietly but firmly. "Their government is in chaos, with fighting in Moscow and who knows what for a budget, and Congress sees it all. We must plan on them participating, but you need to develop a plan that will allow us to continue if they drop out."

"Yes, sir, and I think we can, but the impact could be significant. The sooner we have a decision, the better."

Goldin looked at Shep. "You should assume the Russians are in. The White House has made it perfectly clear that their support depends on it."

Our marching orders were clear, so the team next door doubled down with the Russians to try and answer the technical questions. Having framed the issue to Goldin, I wanted to get a better insight into how much risk might lay in those technical issues.

I requested an hour with the senior Russian engineer to explore how compatible our design philosophies were. I was particularly concerned about their approach to system reliability and how they guaranteed the safety of a human-carrying spaceship. The chief designer for the Khrunichev Design Bureau agreed to meet with me.

After arriving at the offices where the Russians were quartered, I was immediately led to a small office where a grim Russian sat across a small table. The chief designer, who appeared to be in his mid-sixties, was obviously unimpressed to be meeting with yet another NASA manager twenty years his junior. But his grim mood also told me that he suffered no illusions about the future of his country's space program. He knew that they needed the help of these rich Americans if they were to continue.

After a short introduction, I explained the purpose of my visit through the interpreter.

"I am sure you agree that we design and operate our systems differently. Obviously, both approaches work, but I am wondering how hard it would be to get them to work together. My particular concern is what happens when things do not go as planned—when hardware breaks or errors occur. Could you please explain how you design to account for system failures? How do you make your systems reliable? How do you test, and how do you manage your backup systems?"

He seemed surprised by my questions but in a way I could not understand. The surprise lasted only a moment before he began to answer as though he had been thinking along the same lines. He was thorough and addressed each point as a professional working on a common problem. Yet despite his openness, there was something odd about his attitude, as if our conversation was somehow refreshing.

We were both pleased to see that our approach to reliability was similar, but we also knew the devil would be in the details. When he described redundant systems and components, I pulled out a piece of paper and sketched a crude system diagram.

"Do you make these elements interconnected so they can be inserted individually, or do you keep the redundancy at the system level?" I asked, pointing to the system schematic. He took the pen and began to redraw the system.

"We do it both ways, depending on the system and the elements," he said through the interpreter. "Sometimes we must react quickly and switch to a complete backup system. Other times it is better to bypass a component, but that requires more time and care. We must write a special command to make such a change, and it must be the right one. And we do not want more switches than necessary. It can do more harm than good to have everything connected individually. That creates more opportunities for problems."

He was describing the same lessons we had learned and the decisions we had struggled with. He went on to explain that every system was unique and brought its own set of trade-offs. Things that had to function perfectly and react to failures in critically short periods had complete redundant systems that kicked in quickly, sometimes automatically. Other less urgent cases allowed controllers to bypass individual system components. Finding so much commonality in our approaches and experiences built my confidence, and I left our meeting optimistic that we could make this work. He even managed a wan smile as we shook hands and I departed.

Others had told me of difficult meetings with the Russians, during which information had been exchanged only grudgingly and sometimes not at all. That was what I had expected, so I was delighted by our professional discussion.

My thoughts kept going back to the odd reaction I had sensed. As I was preparing to leave the suite, I asked the interpreter about it.

He laughed. "It's simple. You treated him with respect. Most other Americans begin their meetings by lecturing him on spacecraft design as if he were a child or telling him that his program would depend on NASA. Both are extremely insulting to these proud and accomplished men. But from the moment I introduced you, you treated him like a professional colleague working on a common problem. He was relieved and grateful. That was one of the most productive meetings I've been in here."

His words taught me two things: First, mutual respect is the foundation on which trust must be built, and second, the interpreters saw many things we did not.

> **Reliability vs. Redundancy**
> Spacecraft that carry humans must be as reliable as possible. Not only is space inhospitable, but the stressful launch and reentry phases demand that the spacecraft perform nearly flawlessly. However, there are limits to how reliable we can make hardware, and one way to overcome those limits is by installing redundant elements that can immediately take the place of failed hardware. Redundancy can be expensive and heavy, but when done well, it provides the system with the reliability it needs to operate safely.

The discussion with the Russian designer relieved a lot of my concerns about what was possible, but many details still had to be resolved. It was soon agreed that NASA would charter a small group of engineers and operators to travel to Russia to become familiar with the details of how their systems were designed and operated. My old boss, Dennis Webb, would lead the group because of his expertise in the systems we were designing for the American elements.

The second risk, the much more difficult question of whether the Russians would be there when we needed them, would be answered as events unfolded over the long term. In other words, only time would tell. The former Soviet Union no longer existed, and the economic structure that had supported its member countries was gone. It was impossible to tell if anyone could influence the outcome. One of the first groups we sent to Moscow was driven through the city during periods of massive civil unrest, including times when Russian army tanks shelled their own government buildings—an image that did little to build confidence.

We were on the horns of a grand dilemma. President Clinton demanded foreign policy benefits as well as the technical and scientific goals NASA had offered to justify the space station. He wanted to keep Russian aerospace industries busy with us rather than working on new weapons for our enemies. The White House made it clear they would only support the program if we included the Russians.

However, Congress was less supportive, and the fragile state of the Russian government gave them good reason to be skeptical. Several members demanded NASA not create a design that required Russian participation in order for the program to be successful. In the language of engineering, they insisted that we not create a design where the Russians were in the "critical path" of the program, where their failure to perform would render the entire program a failure and our investment wasted.

A traditional design approach would require an either-or decision: They would either be in the design or be excluded. With that option not available, it seemed that we would need a miracle to make both sides happy.

Threading the needle between these positions soon became my job. Shep and Dan Goldin asked me for an approach that would allow the Russians to have a significant role in the program while still allowing it to continue if they were forced or chose to drop out. It was like trying to design an airplane without knowing whether it would have one engine or two. Creating a design that could accommodate both outcomes would cost much more and be more technically challenging. I was being asked to conceive an approach that would let us decide at the very last minute, even after the airplane was already built—perhaps after it was already flying!

NASA is blessed with incredibly creative and hardworking people, and once the problem was laid out, we developed multiple options for how to proceed. My preferred solution was to make the key Russian segment interchangeable with an American-made component that could take its place. In the early days of assembly, that could be the satellite bus that had been included in Option A. If things went well, we would use the Russian module, but if they dropped out, the bus would carry the load—at least for a while.

The detailed work on how this could be done was led by one of the JSC stars from Freedom, Keith Reilly. Keith and his team did the heavy lifting of the system design using my ideas to guide their work. At the same time, I tried to guide NASA management toward better solutions to be considered later.

We briefed Dan Goldin a week after he testified before Congressman James Sensenbrenner, a Republican who was extremely critical of including the Russians. He had savagely attacked NASA and Goldin, but Dan had held his ground and stalled for time while we'd done our work. When I briefed him on our proposed solution, he was ecstatic, literally bouncing in his seat with excitement. He finally had an answer that satisfied both the White House and Congress. We had cleared our biggest political obstacle.

Now, all we had to do was make it work.

9. COST GROWTH

The autumn of 1993 brought an escape from the insanity of Crystal City and a warm welcome from my dear family. After returning to JSC, it was hard to believe how much my life had changed since the beginning of the year. I had gone from being a minor manager in Freedom to one of the leaders in the new Space Station Alpha program. My contributions had been substantial, including the solution to Russian participation, which had led to the White House giving us permission to proceed with the new program. It seemed like the perfect end to a very difficult year.

But the cost had been high. I had been forced to leave MOD because my services had been needed elsewhere. When I talked to Gene Kranz about it upon my return to Houston, he was uncharacteristically ambivalent. Others told me that he thought any decision to leave MOD was a mistake, but he had not stood in my way. I hoped he could tell how difficult the decision had been and how grateful I was for the opportunity and support he had given me.

My family had paid a heavy price as well. Not only had I gone to Crystal City for more than three months, but even after I returned to Houston, the hours were punishing. It was a rare dinner that I shared with my wife and kids. Worse, the stress was causing serious health issues, including breaking three teeth and developing temporomandibular joint problems, leaving me in constant pain. At least for the moment, I was being a lousy husband and father. My dear family, to my undying gratitude, understood that I was doing something important enough to be worth it. I hoped they were right and that their love and patience would be up to the challenge ahead.

The new year brought hope that things might finally settle down. We had hired much of the staff we would need for the program office, including a permanent program manager. We hoped for more reasonable work hours and relief from the political and organizational chaos.

Randy Brinkley was the new program manager. He'd joined NASA a few years before with the expectation of becoming head of the Flight Crew Operations Directorate at JSC, the organization that managed the astronauts and aircraft operations. Someone had promised him that role without considering the views of Acting Deputy Administrator George Abbey. George had built his power base at JSC through the selection of astronauts and their flight assignments. He would not relinquish that power to someone who owed allegiance elsewhere.

As a result, Randy had found himself a brand-new NASA employee without a job. A retired marine corps colonel and fighter pilot, he would not be content with an administrative role. He wanted a position of leadership, and one was soon created for him.

The Hubble Space Telescope was launched with much fanfare in April 1990, only to find that it had a serious flaw in its primary mirror. The science community was devastated that their new telescope was so seriously handicapped, but the talented Hubble team immediately set to work on a solution. They crafted a plan to build corrective optics that would restore its vision and designed a mission to carry out the tricky task. Randy was given the newly created position of mission manager for the repair.

When that mission was successfully completed in the fall of 1993, Randy again found himself in need of a job. Everyone knew that Shep had been struggling for months, and although he was not happy to be replaced, he seemed relieved when Randy was named his replacement. Shep remained as deputy program manager, but the power shift was immediately obvious.

When Randy took control in January 1994, change rolled through the program like a tsunami. He showed no ownership of the program or obligation to those who had been working on it. This probably meant little to the rank and file, but it was a shock to the tight-knit leadership team Shep had assembled, particularly those of us who had been reluctant to join the troubled effort. Randy's dictatorial leadership style and indifference to our views caused our commitment to suffer grievously. Where we had once been a close-knit team committed to each other, we were suddenly burdened with an indifferent boss to whom we felt no loyalty. Although we were still committed to the success of the program, our personal connection was greatly weakened.

This damage was exacerbated when Randy selected someone to be his chief of staff who savored the bureaucratic power of the position. In the past, we senior managers had direct access to Shep and could have frank discussions when needed. Now, the chief of staff was the gatekeeper, limiting our access to Randy and issuing directives without explanation or easy recourse.

This was more than irritating; Randy was making difficult decisions without our background in the issues. Worse, he did so without an obvious process, leaving us with no idea whether we had been heard. His behavior made us feel that we had been relieved of our authority and responsibility. Shep could see it but, when asked, could only give a sheepish shrug, apparently powerless to correct it.

Despite the damage, nobody questioned Randy's intelligence or work ethic. He threw himself into the new job with the intensity that became his trademark. It was common to find multiple emails from him time-stamped at four o'clock in the morning. We appreciated his commitment, but we were deeply concerned that his autocratic style and lack of understanding were doing major damage.

Worse, many of the issues that had threatened the new program were still open but not on his radar. Many, like the political and international issues, required input from every department. Unfortunately, his decisions often seemed to ignore the work we had already done, such as the continuing challenge of balancing the needs of the research community.

Where we had struggled to maximize and balance the station's research capabilities, Randy's approach was to personally decide what would be provided, granting some requests while denying others. Although we assumed he had his reasons, his decisions appeared to be either arbitrary or based on favoritism, both equally inappropriate.

Cost and schedule remained among our most immediate issues. One strict boundary was a total budget through development of $17.4 billion, including a healthy reserve, which the transition team had vigorously enforced. Under Randy's leadership, our precious budget and schedule reserves were disappearing rapidly in a blizzard of new changes. Many of his decisions were made before the impacts of earlier changes were understood, making the budget and schedule rapidly moving targets based on immature estimates. They both kept slipping as our confidence kept eroding, creating ample fodder for our critics.

Having been charged with controlling the program risk, I was alarmed. Many more programs are canceled because of cost and schedule growth than technical failures, demanding close attention to their performance. But Randy had no respect for risk management of any sort, and his lack of interest, coupled with the blizzard of changes coming from his office, made my job impossible.

When he would not listen to me, I tried to work through the lower levels of the program, but his management style made my efforts futile. Worse, his dismissive attitude toward risk quickly spread to lower-level managers. I was horrified to see a repeat of the behavior that had gotten Freedom into so much trouble.

Unable to do my job, I went to see Randy. I carefully told him that his management style had changed the role of my office to where it was no longer a good fit for me. Then, I asked to fill the vacant seat as manager of the top program-level integration forum. Although I knew he would retain his death grip on the decisions, the new role would give me better access to the overall health of the program and hopefully allow me to resolve some of the most important issues. I also volunteered to work more closely with the prime contractor, Boeing.

Randy agreed.

Boeing had been chosen for this important role before Randy became program manager. They were selected because of their success in supporting many different agencies. Their new program manager, Larry Winslow, had become a star for his work on the F-22 program. Now, he had been told by Dan Goldin that he was responsible for the success of our program. He eagerly accepted that responsibility and used it to support his strong leadership style. Most of us were delighted to see his commitment and accountability, but the blanket authority he assumed for Boeing was creating as many problems as it solved.

A tall, imposing, and extremely bright gentleman, Larry was not shy in exercising what he felt was his authority to fulfill his responsibility. He recognized the challenges he faced and was clearly committed to making the program successful. He also realized the urgency of the work, especially getting the requirements finalized. But he was new to spaceflight and had no understanding of the business or the details of the work. Many NASA folks had expressed concern about his lack of space background, but we were told to support him and make Boeing successful. Most of the NASA team and I did exactly that.

Randy Brinkley did not. He often complained that their work was intruding into NASA's areas of responsibility and opined that Boeing was getting "too big for their britches." While I agreed that the confused roles and responsibilities were a problem, I saw it as NASA's failure to lead rather than Boeing's aggression. Seeing a crisis brewing, I urged him to invite Larry to his office to resolve these concerns. After all, we had given the senior Boeing team offices on the same floor of JSC's Building 4 South as the civil servants. The distance between the program managers' offices was less than two hundred feet, so coordination should have been a snap. But because of the hostility coming from Randy and his chief of staff, the distance might as well have been two thousand miles.

Since Randy showed no interest in resolving the issue, I volunteered to be his proxy. Having spent a lot of time with Boeing, I respected Larry and felt he reciprocated. Randy quickly agreed to my plan, no doubt to get me off his back, and I immediately headed down the hall.

As a frequent visitor to the Boeing area, I was comfortable dropping in on their management team. I stuck my head in Larry's office and asked when would be a good time to talk about roles and responsibilities. With his usual directness, Larry said, "There's no time like the present. Pull up a chair!"

Knowing his preference for tackling issues directly, I had already thought out the approach I wanted to take.

"Thanks for making time, Larry," I began. "I am very impressed with the team you have assembled and the quality of their work, but I also see issues brewing over authority and responsibility between your team and NASA."

"I agree," he said succinctly. One of the things I liked most about him was that he spoke directly and without ambiguity.

"Then I am sure you agree that it is important to resolve those issues before they become a distraction." He nodded as I went on. "Could you tell me exactly what you see as the scope of your responsibility for this program?"

"That's easy. Dan Goldin has made it perfectly clear that Boeing is responsible for everything. The whole enchilada." He smiled broadly, his pride clearly visible but not offensive. He knew this was a big challenge, and he was very proud to have been entrusted with it.

Nodding, I said, "I'm delighted to see you accept the challenge, but help me understand a little more of the detail. Obviously, you have full responsibility for the design and development of the major US elements that will be flown in space. But how do you see your responsibility for mission operations?"

"We don't do operations. We will provide the technical documentation and sustaining engineering data for the elements," he said confidently. "Actual mission operations are NASA's responsibility."

"I agree," I said, nodding. It was nice to start out with areas of clear agreement. "Let's take on something a little harder. What is Boeing's responsibility for the government-furnished equipment?"

Again, he crisply described the roles, showing that he had thought about the issue and discussed it with his team.

"We establish the requirements and integrate the equipment into the system, but we are not responsible for getting it delivered or how it performs. Because we do not hold the contracts, we can only assume it will do what the government claims it will. We cannot be responsible for what we don't control." His steady gaze made it clear he was confident in the position.

Going on, I said, "That sounds right to me too. What about the international partners? What responsibility do you have for their contributions to the program?"

"Same thing. Boeing is the integrator, so we are responsible for ensuring that the requirements have been established to make the elements work together correctly, but we are not party to the treaties. That is a government function, so it is up to NASA to ensure that the parties perform as promised."

"It's hard to argue with that," I said. "What about considering the risk of whether the partners deliver as promised?"

"That is NASA's job, although we will support."

As we talked, I took out a piece of paper and sketched a series of vertical bars representing the many activities across the program. Capturing what we discussed, I drew a horizontal line across each bar to approximate the split of what we had agreed were our relative contributions. This created a map of roles and responsibilities that we could talk about and share. Over the next hour, we covered the whole program, and as the details of the picture were filled in, I could see that Larry was surprised that the story it told was quite different from what he had believed only minutes before. The new picture clearly showed that Boeing had broad responsibility for analyses and plans but that

NASA held many critical roles. The map even showed how the two organizations would have to interact.

"Wow! That was illuminating!" The surprise and relief in Larry's voice were clear. "Thanks for taking this on, Jim. It was greatly needed and will help us in so many ways. It will be absolutely priceless as we finalize the contract!"

"My pleasure, Larry. I'll turn it into a PowerPoint chart and get it back to you for comment before the end of the day. I am sure there are tweaks to be made, but it's clear we are on the right track, and having it written down will make life a whole lot easier."

"You got that right," he said with a laugh. "It's a wonder we got this far without it!"

Returning to Randy's office, I was eager to show him our product. He listened as I explained the conversation but seemed little interested. He said, "Put it together, and I will let you know how I feel about it then."

An hour later, I took the cleaned-up chart down to Larry, who called in his deputy, Doug Stone, and his business manager, Scott Carson. All were excited to see it and, after a few small tweaks, agreed that it correctly described the basics of how we should proceed.

The next morning, I took the finished chart to Randy and reviewed it once more. He mumbled something unintelligible and put it in his desk drawer. I was surprised and disappointed that he did not seize this opportunity to achieve closure on this critical issue. It would be a few weeks before I understood why.

Everyone knew that Randy and Larry did not get along. The tension between them was a major reason for my push to clarify roles and responsibilities. We needed a fully productive team, and I had hoped that this effort would clear the way for the hard work ahead.

The workload was already tremendous, with many people working eighty hours or more per week. Most people can do this for short periods, but when it goes on for months or years at a time, it becomes exhausting. Since we knew we had many years of this ahead of us, we made a point of having regular celebrations to relieve the stress, build team spirit, and celebrate our accomplishments. Toward the end of 1994, we had such a celebration at Clear Lake Park, a few blocks from JSC. It was a beautiful evening with many reasons to celebrate our progress. I looked forward to sharing some stress-free time with my new friends and colleagues.

As soon as I entered the park, I could tell something was wrong. My Boeing colleagues were tight-lipped and grim, while Randy Brinkley and his chief of staff seemed unusually happy. When Randy stood up to address the crowd, he spoke eloquently of Larry Winslow's many contributions to the program before announcing that Larry would be returning to Seattle to take on "new challenges." Larry was gracious and talked about his new job as a step up, but his demeanor made it clear that he had been forced out. Despite Randy's words of gratitude and regret for Larry's departure, nobody bought it, and the effect on the Boeing team was obvious.

I felt terrible that Randy had been so underhanded. Because of my close working relationship with Boeing, my colleagues pulled no punches in telling me their views. Even Larry's deputy, Doug Stone, who was being promoted to the new Boeing program manager, was furious that Larry had been removed over the objections of the Boeing team.

The chill in the relationship was felt immediately, causing me to put even more effort into enabling communication between the NASA and Boeing teams. Both Doug Stone and Scott Carson, who had moved up to become Doug's new deputy, knew that I was deeply embarrassed by Randy's act. They made a point of being completely open with me, giving me full access to all of their management forums. They even seated me between them during their weekly program review, making it clear that they would hide nothing from NASA or me. For my part, I publicly applauded their close attention to the work, complimented their commitment at every opportunity, and redoubled my efforts to improve communications with the NASA team. They gave me full insight into their work and complete confidence that they were doing their best to resolve the problem areas. When I asked if other NASA folks could attend their weekly review, they replied as one: "Hell yes!"

To my great embarrassment, no other senior NASA manager accepted their invitation. Worse, some of those same NASA leaders who refused to attend were openly skeptical in our NASA management meetings. Often, when I reported what I had observed, Randy's chief of staff would cynically dismiss my inputs as "more of the Boeing Kool-Aid." Sometimes, he even did this in public meetings, exacerbating the already-frosty relationship with our contractor. For the life of me, I could not understand how he thought it was helpful to undermine our relationship with our most important teammate.

As the spring of 1995 began, the cost and schedule problems continued to mount. Under enormous pressure, NASA and Boeing had agreed to the terms of the prime contract literally minutes before midnight on December 31, 1994—the deadline NASA Administrator Goldin had established. As a practical matter, the "agreement" was really just Boeing accepting the top-down cost as dictated by NASA, not a bottom-up budget based on realistic costs from subcontractors. This meant it was a target, not a commitment, and was likely to fall apart.

As if to reinforce this, a senior Boeing cost analyst pulled me aside and warned, "This agreement is setting us up to fail. We do not have agreement on even the basic elements, with a ton of work yet to be defined. The surest path to a big cost overrun is to start out without enough funding to cover the baseline." Although he was preaching to the choir with me, events would prove just how right he was.

By March 1995, I could see the pattern of cost overruns bearing down on us. We still did not have a workable budget, and NASA Headquarters was already pressing for new requirements that the program could not possibly afford. As the program manager, Randy tried to juggle people's expectations without admitting the gap, but when I sketched out the likely outcome, I came up with an overrun of at least $3 billion. One sunny Friday morning, I took my analysis to Randy and Shep along with my recommendations on how to fix it.

Randy's office in the southwest corner of the fourth floor offered a beautiful view of the gorgeous spring day. The brilliant white clouds and deep blue sky lifted my spirits and made the difficult conversation a bit easier. While the others listened quietly, I explained my analysis of how the changes were eating up our reserves and the likely outcome if no action was taken. I then showed the steps I thought would frame the balanced program we could afford.

My spirits collapsed as Randy dismissed my analysis and recommendations out of hand. He tapped his forehead with his right index finger and said, "I have it all planned out. I don't need your help."

I was momentarily speechless. Although I had expected him to challenge my input, I was stunned by his total dismissal. Somehow, I managed to keep my cool as I tried again to make clear how serious I thought the problem had become. When he again dismissed my input, I could not help but say what my two years of total commitment demanded in words straight from my heart.

"Randy, long before you came on board, I told the Vest Committee that we would bring this program in on cost and schedule. They were extremely skeptical that NASA would live up to its agreement. This conversation seems to support their view. If that is your whole response to what I have reported, I'm not sure I can remain a part of this program."

Randy shrugged. "You need to do what you need to do. I'll help you find another job if that's what you want."

Struggling to keep my composure, I excused myself and returned to my office. Although the personal rejection stung, I was furious to see that the whole program, built on so much hard work and sacrifice, was being threatened this way. I had not expected Randy to accept my recommendations outright, but I had expected him to respect my concerns and ask for additional data and options. That is all I expected from any serious professional. His decision to simply ignore me and my analysis left my confidence deeply shaken. From my first day as a member of the transition team in Crystal City, I had committed myself totally to the program's success, and it had never occurred to me that I would leave it less than two years later because I thought it was untenable. I was heartbroken.

It was time to think about my options.

After taking a few minutes to lick my wounds, I walked down the hall to join the weekly Boeing review. Seated in my usual place between Doug and Scott, I knew they could see that I was distressed, but they did not ask why, and the busy agenda soon distracted us.

About midafternoon, Randy's secretary came in and handed me a slip of paper. Randy wanted me to join him in his office as soon as possible. I excused myself and said I'd be right back.

Walking down the hall, I hoped he had decided that my concerns deserved to be considered more thoroughly. Maybe he would even welcome my help in resolving them.

Instead, he took the conversation in a new direction.

"Jim, I've got a problem that I think you can help me with."

"Happy to do anything I can, Randy."

"There's a guy down at KSC who is ignoring the program position about logistics. He's pursuing his own approach without regard to the program plan. That is absolutely unacceptable, and since I know you have a background in the area, how about going down and fixing this?"

"I'm happy to help," I said. "Am I empowered to negotiate a solution?"

"Do whatever you need to do to get the program back on track."

Walking back to the Boeing review, I felt as though I was breathing my first breath of fresh air in months. An hour before, Randy had totally dismissed my arguments on the cost overruns, but now he had asked me to fix something and given me the authority to do so. The difference between feeling impotent as I watched cost overruns accumulate versus being empowered to resolve this problem, small as it was, was breathtaking. It was time to find a job where I could accomplish useful work.

The issue at KSC in Florida came from a new manager who did not appreciate the size of the program and the implications of his decisions. Despite being exceptionally intelligent (proudly broadcasting his membership in Mensa), he was very shortsighted. I listened carefully to his proposals and helped him save face by incorporating a few of his suggestions, but ultimately, we returned to the program plan because it was the right approach.

Returning to Houston the next day, I reported my actions to Randy, who was pleased. Once back in my office, I contemplated where I should go next. My leadership roles in the space station program had given me contacts in every organization at JSC, and I could probably find a position in almost any of them. For some reason, the one that struck me as most appropriate was the Shuttle-Mir program, which had grown to become the first phase of the space station program. It would bridge the American space shuttle experience and Russian history on the Mir and Salyut stations. Its work would flow into what would eventually be called the International Space Station (ISS), and I hoped it would someday give me the opportunity to rejoin that program with new and valuable insights.

Picking up the phone, I called the Shuttle-Mir program manager, Tommy Holloway. We had become acquainted as a result of some EVA issues arising from earlier shuttle missions, and he had even tried to hire me for a new job he was creating to deal with them. While I had been interested, the offer had become moot when I'd been chosen for the space station transition team. Now I hoped that he would not hold that decision against me.

Events would once again prove the wisdom of this old saying: "Be careful what you wish for."

10. THE MOVE TO PHASE 1

The fall of the Soviet Union in December 1991 marked a turning point in world history. The Cold War, which had threatened our very existence for over four decades, was suddenly no more, and the enormous Communist bloc was replaced by a much looser Commonwealth of Independent States. Those of us who had lived in the shadow of the Cold War all our lives, and especially those who had been on the tip of the spear defending against nuclear war, were deeply relieved.

NASA was soon awash with ideas, large and small, for how the agency should begin cooperative work with its Russian counterpart, the Russian Space Agency, or RSA. By the summer of 1992, work had begun on a successor to the Apollo-Soyuz Test Project where NASA would fly a Russian cosmonaut on the space shuttle and an American on the Russian space station Mir, with each country participating at its own expense. This Shuttle-Mir program was a very modest proposal, but we all hoped it would be the beginning of something more.

The decision to create the Shuttle-Mir program had been made before the space station redesign even began. When the redesign was approved, the American and Russian teams were asked to merge their programs while using Shuttle-Mir to gain experience working with existing vehicles before taking on the big problems with the new station. The original plan for a single flight on each other's vehicles was expanded to allow up to ten missions to Mir, which would form Phase 1 of the ISS program. Together, the programs were called the

Shuttle-Mir Phase 1 Program, and my decision to join the effort came at exactly the right moment.

Tommy Holloway was now the program manager for this combined effort, and my call to him led to an invitation to come by for a chat. Our prior contacts had shown him to be an old-school manager who held his cards very close to his vest, so I was not surprised when he was initially noncommittal about my joining the effort.

He described an office with no organizational structure—everyone was simply on his staff. While he had a vacant slot from a recent retirement, he was not sure how or even whether to fill it. He did admit that there would be lots of work soon enough, and with considerable misgivings, I decided to join his team.

Once on board, I found that his small team was extremely busy. Tommy sat at the center of everything and made most decisions personally. His deputy, Frank Culbertson, was an energetic navy F-14 pilot-turned-astronaut with a longtime interest in Russia. At first, I rarely saw Frank because of his many trips to Russia. His growing personal relationship with the Russian leaders would help us through the many rough spots to come.

Leading the crew planning and coordination was Don Puddy, a veteran of Apollo who had held many responsible positions, from flight director to head of flight crew operations. Don provided a quiet but forceful presence in the office. Although he would soon retire, his experience added credibility to this newborn program.

Kathy Leary had come from the shuttle program to integrate our missions to the Mir. Kathy had a sharp wit, a biting tongue, tons of energy, and a drive to get the job done as well as humanly possible. Her close contacts and deep understanding of the space shuttle processes were invaluable.

Charlie Brown was a quiet gentleman who had come over from the MOD training organization to negotiate the training requirements for our crew members going to the Mir. He was immediately liked by his Russian counterparts, which helped us through many difficult spots.

Travis Brice was another unassuming character whose quiet demeanor concealed a long history across many areas at JSC. He had come to the program early on to help with contracts and logistics needed to keep the program moving.

George Nield was as new to the effort as I was, coming from the same part of the Space Shuttle program as Kathy. Where Kathy managed our internal requirements, George represented us at Space Shuttle program meetings.

Dave Hansen had spent a year living in Moscow and wanted to help bring our countries together. His Russian language skills and familiarity with Moscow helped us support our folks who stayed there.

Two office support contractors worked hard to keep the paperwork flowing so that we could execute the mission, and our two office administrative assistants, Jesse Gilmore and Lindy Fortenberry, managed an enormous meeting and travel schedule.

Although not part of the Shuttle-Mir/Phase 1 program office, Rick Nygren and his life sciences team were a critical extension of it. Their original job was to manage the research to be conducted on Mir, but they would soon assume a far greater role.

Curiously missing was any representative from MOD, but given Tommy's background in that organization, I assumed he had it covered through some other process. My first impression was that MOD only supported the shuttle docking missions, leaving the long-duration mission strictly to the Russians. This would turn out to be both a conscious decision and a serious error.

My first day in the office was April 1, 1995, accompanied by the inevitable jokes about April Fools' Day. For me, there seemed little to laugh about, as it felt that my career had taken a giant step backward. Just a few days before, I had been a senior leader in one of NASA's largest programs, and now I was a junior staffer in one of the smallest.

Almost immediately, I was struck by an odd fact: While the staff was building plans for future crews and briefing senior management, there was almost nothing said about the mission underway. There was no sign that just over two weeks prior, NASA had launched its first American to the Mir.

Norm Thagard was that first American to visit Mir. He was a highly accomplished individual—a medical doctor, engineer, marine fighter pilot, and astronaut—and had flown to the Mir aboard the Soyuz capsule along with two cosmonauts to begin a three-month stay. Although I had not met Norm, I was confident that his experience as a marine aviator had prepared him to deal with almost anything. I would soon realize that his selection to be the first American on Mir was an incredibly fortunate choice.

For the first few days in the office, the only reference I could detect to the ongoing mission was an occasional mention of the Mission Management Team (MMT) meeting. I was familiar with the Mission Management Team as the means to oversee other missions, and I was curious to see how it was working for this program.

Clearly, the MMT for Mir missions would be different from the shuttle version, but I had no idea what to expect. At the appointed hour, I joined Tommy, the office staff, and Rick Nygren's science team in a small conference room. We communicated via telecon with the American flight surgeon located at Russian mission control just outside Moscow.

Dr. Mike Barratt was the flight surgeon supporting Norm, and he played a critical role in Norm's mission. He had accompanied Norm through much of his training and had supported him with countless aspects of his life in Russia. Now, he was the only American the Russians allowed to communicate with Norm each day. The few precious minutes allocated for their daily communications went quickly as they had to support Norm's research, conduct public affairs events, and keep Norm in touch with his family.

The NASA team had not appreciated how few those communication opportunities would be. Unlike the space shuttle, which used relay satellites to provide nearly continuous communications, the Russians relied on fixed ground sites. A typical day held about a dozen short communication opportunities as Mir passed within range of a site, and those few minutes had to be shared among all users. Since the Russians had countless higher-priority issues, Norm got very little of the time. The limited communication, together with the inevitable misunderstandings, schedule changes, and problems with the hardware, created chaos. When something was missed, it could take hours or days before the next opportunity to discuss it.

Our team in the Russian mission control center north of Moscow was struggling as well. They were few in number and worked long hours under difficult conditions. Nothing could be taken for granted, and it was even difficult to make routine telephone calls back to the US.

My first Shuttle-Mir MMT telecon was a disaster; the audio quality was poor, there were fre-

> ### MMT
>
> NASA human missions are managed through a two-part process. First, they are planned in great detail, with every item and contingency considered, flight rules established, and final plans approved by management. Once the mission is underway, the Mission Management Team (MMT) reviews its progress, and anything that deviates from that plan or triggers a flight rule is discussed. For shuttle flights, preflight planning is largely sufficient, with the MMT mostly concurring on the daily plans. However, the far greater length of station missions significantly increased the MMT's role, often requiring difficult decisions on a daily basis.

quent interruptions, and nobody seemed to know who was doing what. Norm's research program was in chaos because his support equipment was not working, the Russians could or would not help, and our team was powerless. The confusion and impotence were palpable, as were the anger and embarrassment of the team in Russia, yet nobody in Houston stepped forward to fix these problems.

The responsibility for the meeting fell squarely on the program office, and I was stunned. The weak leadership, technical challenges, and difficult relationship with the Russians portended nothing but trouble in our future. I felt panic begin to rise, even as I tried to limit my focus to Norm's safety and our modest research program. Those were responsibilities enough.

After the MMT meeting, as Tommy and I drove back to our office, I pondered what to do about the fiasco I had just witnessed. Tommy was highly regarded across the center, and I did not want to be disrespectful, but it was clear that we could not continue to operate this way. Somehow, we needed to assemble a stronger team, develop a better plan, and negotiate better support from the Russians.

"That was not good," I said quietly.

Tommy said nothing.

"What is our agreement on roles and responsibilities for Norm's mission?" I asked.

"All we are responsible for is the research program," he said quietly. "The Russians promised to take care of everything else, like they have for other visiting cosmonauts."

During the Soviet era, the Mir program had often flown visiting crew members from other countries, usually satellites of the USSR. However, those had always been short visits done for publicity purposes, not long research missions, and little support had been required or provided. Once back at the office, I resolved to take a critical look at the planning for Norm's mission.

It was quickly obvious that the problem lay with NASA's lack of support for mission operations. All the functions that MOD provides for a NASA mission, from developing the mission plans to leading the team through execution, were missing. Norm was trained by NASA people on his experiments, but everything else was left to the Russians.

When asked about this gap, Tommy pointed to a bilateral agreement that explicitly said we could not participate in the operation of the Mir. Our total responsibility was training Norm on his experiments and taking care of his personal needs. Although I was shocked by the passivity of the agreement, it might have been acceptable if Norm were going to be the only American to fly on Mir. But he was not, and I could not imagine sending four more Americans to suffer the same fate.

"Didn't anybody think about how big a leap NASA was making from the Russian experience?" I asked my colleagues when we got back to the office. "Guest cosmonauts only stay for a week, but our folks will be there for months. And we have lots of research hardware coming on the *Spektr* and *Priroda* modules that will require engineering and operational support, especially in the beginning. How can the NASA team possibly do its job?"

"Not our spaceship," Kathy Leary replied. "The Russians are responsible for mission operations, as they constantly remind us. Tommy and Ryumin agreed that we would only manage our science program and leave the rest to them. We have to fit into the resources they give us."

Valery Ryumin was the tough, burly, former Russian army tank commander-turned-cosmonaut who headed the Mir program. I had met him in our early meetings with the Russians but had not worked with him directly. Listening to him on the weekly telecon that he had with our program office gave me little optimism that things would improve. Like most Russian leaders of his era, he had a loud, bellicose voice and showed little interest in listening to our concerns. He treated us as minor irritations, and Tommy did not press him.

"Our whole program is dependent on their actions," I said. "We need access to facilities, power to run experiments, and communications time to coordinate our work. How are we handling these issues?"

"Good question," Kathy replied.

Tommy had a different view.

"It's their ship, and we can't run it for them. They made an agreement to support our program, and that is that. Ryumin insisted that we were no better than the other guests they have had on Mir and was not about to let us into their business. Even if I complained, it wouldn't make any difference."

I was stunned by his passivity.

"But it's not working," I said. "That MMT was a disaster. Poor Norm is struggling to get anything done, and we are doing damn near nothing to help him. And if the Russians have a serious problem, we are completely out of the loop, incapable of managing the mission."

"That is what Goldin signed us up for," Tommy said angrily. "Talk to him about it."

Although Tommy was being facetious about calling Dan Goldin, I briefly considered doing just that. My experience had fueled a powerful sense of responsibility for our programs and the courage to stand up and speak when it was appropriate.

If and when it became appropriate to call Goldin, I would do so, but for now, I decided to wait. Having barely joined the office, I knew almost nothing about what was going on and why. What passed for evidence of the program's management approach, as bad as it was, was pretty thin. Things needed to play out a while before calling the top brass.

The downward trend continued. Over the next few weeks, I found it increasingly painful to watch Norm's mission devolve into utter frustration. One of his primary research tasks required him to collect blood and saliva samples that could be analyzed later for data on how his body was affected by his long stay in space. Those specimens needed to be kept frozen to preserve their scientific value, and NASA was counting on the use of a freezer developed by the European Space Agency (ESA) for their Euromir 94 mission. They had assured NASA it would support our needs despite not having been used for months.

They were wrong. Every day, Norm would awake to find the freezer's heat exchanger thoroughly frosted over and its insides barely cool. And every day, Norm would spend hours removing the samples and wrapping them with whatever insulation was available while he struggled to remove the frost. Besides being tedious and frustrating, it was an enormous waste of time, and the samples were eventually ruined. Through it all, the Russians watched passively, offering little information and even less support.

To his enormous credit, Norm quietly did his job, making it his mission to protect the samples, even if it meant defrosting the freezer every day. He simply put his head down and did his very best.

It was agonizing to watch that happen, but what really bothered me was the way the Phase 1 team, from Tommy on down, stood by while the problems unfolded. When I asked about the problems with the freezer, I learned that NASA had almost no data on the unit and no plan to obtain it.

I was becoming apoplectic; this was a complete reversal from the way NASA conducts its own missions, with experts and hardware standing by to assist. During a space shuttle flight, mission control swarms with experts in every system. We had nothing, and nobody seemed to care!

The only resources for troubleshooting the freezer were in France, so I asked if we could have someone visit the factory where it was made to learn about its operation. Tommy agreed that that was a good step, and one of the support engineers from Moscow made the trip. While that gave us a little help, the freezer never performed well.

After I had been in the office for about a month, when it became inevitable that the samples would all be ruined, I had had enough. It was time to speak up.

"Tommy, this is no way to run a mission. Whatever agreement you have with Ryumin, we have *got* to support our crews better than this."

Tommy looked down at his notebook, frowned, and bit his tongue. Although he obviously did not appreciate my complaints, he had to agree. And while I was very uncomfortable pushing my relationship with my new boss, it was necessary, and I think he knew it.

I went on. "We need what MOD does for every shuttle mission. We need their planners and trainers as a minimum, and some leadership from the flight directors wouldn't hurt!"

Tommy shook his head, refusing to even discuss it.

"That's impossible," was all he would say.

His attitude reminded me of times when I had touched on an old family issue, like the escapades of a crazy uncle, that nobody wanted to air out. Everyone does their best to ignore the reality, even as they know they should take action.

Eventually, as the mess continued to grow and I continued to fuss, Tommy called a meeting with the flight directors who supported our shuttle missions to Mir. Assembled in the conference room with Tommy and me were Rick Nygren; Bob Castle, my former host in the Flight Director's Office; Phil Engelauf, another flight director I knew well; and Lee Briscoe, the head of the Flight Director's Office.

Tommy began the meeting by laying out the problems we were having, but instead of asking for help, he coyly hinted that any support MOD wanted to provide would be accepted. The flight directors were not biting.

"That is your problem," said Briscoe. "You made this mess, so you can find your own way out!"

Having briefly worked for Lee during my detail in the Flight Director's Office, I was not surprised by his parochialism. He had a very narrow vision of his responsibilities. What did surprise me was that my former colleagues were blind to the fact that our mission was NASA's mission, and our failure would be NASA's failure. I tried to make that point, but the anger and hard feelings overwhelmed everything else. The meeting ended with hardened feelings on both sides and no progress.

After the flight directors left, Tommy turned to Rick Nygren.

"I know this is not what you signed up for, but since you own all the details, I need your team to put together a plan to run the operations on Mir. You're all I've got to make this work."

Rick responded in his usual matter-of-fact manner.

"I hear you, and I don't disagree, but I don't know how I can do that. I don't have enough people, or the right people. I don't have the right agreements with the Russians, and you know they don't want us in their knickers. We have no leverage over them."

Tommy simply said, "This is going to be ugly, but we have no choice. Put together the best plan you can, and I will see what I can negotiate with Ryumin."

Rick knew exactly what was on the line. Norm's mission was the first of five planned and the simplest. Later missions would need far better planning, training, communications, and resources. Rick would eventually put together an effective NASA team, but it took time to work it out, and Russian cooperation would always be grudging. One thing became clear—the Phase 1 Program would have been impossible without the extraordinary efforts of Rick and his team.

In stark contrast to the meeting about working on Mir, when the time came for the shuttle to make its first docking with Mir, MOD and its flight directors did their usual great job. Hoot Gibson and his crew on STS-71 got the mission done with astonishing precision, making contact within a few seconds of the planned time and setting a bar that other crews had to work hard to meet. They carried two cosmonauts to the station and returned Norm and his two crewmates. The mission also gave us an opportunity to show off the flexibility of the shuttle to accommodate late changes to the mission.

The *Spektr* module had been launched and docked with Mir a few weeks before STS-71. This major research module carried NASA research hardware and four large solar arrays to increase the electrical power on the station. However, when the solar arrays were deployed after launch, one of them did not extend completely. A launch restraint failed to fully release, and the crew needed to conduct a special EVA to cut it loose.

At the next weekly telecon, Ryumin asked if we would add a tool to the shuttle manifest to cut the launch restraint, and we agreed. Although the Russians had been developing a tool for this task, our engineers could not ignore the challenge. They quickly found a commercially available ratcheting cutter designed to cut steering wheels for rescuing people from automobile wrecks. They bought one, made a few changes for EVA, and conducted a quick test in a thermal vacuum chamber. It was a simple, elegant solution.

A few days before the STS-71 mission, the Russian managers came to Houston with their tool, and our engineers provided ours. We met in our fifth-floor conference room to test the two tools side by side. Engineering provided a piece of aluminum tubing similar to the launch restraint that the two tools could cut. The Russian tool did the job just fine, but the American one was much faster and easier. The Russian managers admitted that our tool was better, but they asked us to fly both just in case, which we did. A few days after the shuttle had left Mir, the cosmonauts used the American tool successfully, restoring Mir's power generation capabilities.

The same mission also provided some delightful examples of the human side of our work. My favorite anecdote concerns the Russian cosmonauts' arrival at Kennedy Space Center on the morning of July 7, 1995. The smooth shuttle landing capped a brilliantly successful mission, but in the rush, the Russians had forgotten to bring passports for the cosmonauts. The news that two "space aliens" had landed in Florida caused a short crisis with US Customs, but the bureaucracy managed to accommodate the oversight with grace and humor. It was a good day for NASA, the US, and the space program.

A few days after returning to Houston, Norm came by to debrief our team. We were extremely interested in hearing the details of his experiences working within the Russian system, but we were also dreading his feedback on our pathetic support of him.

This debrief was one of many opportunities that proved Norm had been the right man for the job. He was completely open about his experiences—what had worked, what had not, and what he thought would be needed for future crews. He took pains to describe the cultural isolation he'd felt and how it had affected his mission. For example, his Russian language skills, which met Russian standards for technical work, had been inadequate for sharing more human experiences with his crewmates. This had added to his sense of isolation and made his stay more difficult.

I was delighted at his openness and humbled by his professionalism. Despite all that he had gone through, including the many foolish and parochial decisions that had made his stay difficult, he said not a word of direct criticism. He always focused on what had gone well and what had not without attacking those who had made the decisions. He'd simply done his job as well as he possibly could have.

This was proof positive that our program was doing its job. The space station program needed the Russian and American teams to work together, and this was the ugly process that taught us what would be required. Whatever niceties were said in public, our internal review was brutal, and we looked for opportunities to do much better in the future.

It was time for Shuttle-Mir to become Phase 1 of what would eventually be the International Space Station.

11. LEARNING TO WORK TOGETHER

The STS-71 mission marked another transition for me. In the past, I had been in a minor support role for space shuttle missions, but this time, I was often the decision-maker for the Shuttle-Mir program. With Tommy and his deputy, Frank, handling issues for later missions, I was the senior member of the management team present in the control center. The decisions were modest because everything went well, but it was training camp for what was to come.

The end of Norm Thagard's mission did not signal a slowdown in our activities. Although we had about nine months before the next astronaut would go to Mir, we needed every minute of that time to review what had happened with Norm and explore ways to improve. Even though Norm had been incredibly gentle with his criticism, the message was loud and clear: We had fallen far short in our support of him and the research program. It was time for a change.

"Tommy, we have got to improve our planning and oversight of the missions and especially our presence in Moscow," I said. "We had almost no insight into what was going on and why. It is only a matter of time before something serious goes wrong while we are powerless to deal with it."

Although I was prepared for a fight, Holloway did not argue.

"What do you recommend?" he asked quietly.

"We need someone in Russia full time to represent this office, someone who really understands operations and the details of the program. And it needs to be someone senior whom the Russians will respect and keep in the loop on operational decisions and technical issues."

NASA already had a deputy for operations in Russia (DOR)—in Star City, more formally known as the Gagarin Cosmonaut Training Center (GCTC). The DOR was an astronaut who served as our representative and oversaw support for the crews in training. There were plenty of issues to solve in that domain without worrying about mission operations. And Star City was both remote from and a totally separate organization from mission control, which was located in Kaliningrad, soon to be renamed Korolev.

Tommy sat quietly, looking at his notebook, then looked up and asked, "Are you willing to spend three years in Russia?"

That stopped me in my tracks. It had never occurred to me that he would consider me a candidate for the job. Although I had been a leader in the creation of the new station program, I had little relationship with this group of Russians and less evidence that they would accept me as the senior leader we needed. I also knew that they had little patience for those who did not know the business, and although I was well-versed in the US system, I knew very little about theirs.

I was also concerned that the turmoil of the last few years had been hard on my family. Ever since the kickoff of the redesign in early 1993, I had been working very long hours and coming home exhausted and out of sorts. My family deserved better than what I was giving them, and hauling them off to Moscow did not sound like a step forward. Still, I knew this was important.

"I would consider it," I said carefully, "but I don't think I am the right guy. They really respect gray-haired engineers with lots of experience in the trenches. We need to figure out exactly what we need, then find the right person to do it."

Tommy nodded quietly, and I could see his mind was hard at work looking for an answer.

Closer to home, I pushed for better stewardship of the individual crews and their missions. For the space shuttle, NASA had a huge cadre of people to support the crew with everything they needed. In Russia, the astronauts were almost completely on their own, and they were barely able to keep up.

I imagined the ideal solution would be a team that could train the crew on the research program and then support them through their time on Mir. I also imagined a new relationship with the Russians that would allow us to participate in decisions that affected our program before they caused problems. On Norm's mission, we were left to react to the changes, often with major impacts on our plans. In addition to being frustrating, it left our research program in tatters.

After many long debates, we decided to select a series of operations leads—which we called ops leads—each of whom would be dedicated to a specific astronaut for their mission on Mir. They would work and travel with the astronaut to understand everything about the mission, from training to execution. With the help of a few colleagues, including the flight surgeons who continued to support our crews, they would coordinate with the Russians, support the research program, and be the crew's prime liaison on the planet. It had the potential to be a great job, and I hoped that many people would volunteer.

As with any pioneering effort, the first one would be the most important. We needed someone with both the requisite experience and the willingness to put his or her career on the line.

"How about Bill Gerstenmaier?" I asked Tommy. This was the individual who had impressed me during our work on Freedom assembly. I had not seen him in several years but had talked to him at the beginning of the new space station program to see if he would join my team. At the time, he'd been pursuing his doctorate in aerospace engineering at Purdue University and had not been interested. Now he was back at JSC, and I hoped he would see this opportunity and the challenge it offered.

"Bill would be a good fit," Tommy said, nodding slowly. "Do you think he would do it?"

"Only one way to find out," I said, "but I think he might. He likes a challenge."

Tommy called him, and Bill agreed. Getting him for the first ops lead was a great step forward. We didn't have to spell out the problem for him. He quickly got the picture of what we needed and set out to make it happen. Although the model he created was not exactly what we wanted, it got the ball rolling in the right direction. We would make adjustments later.

The harder part was finding three more leads to support the remaining crews. MOD had the people with the skills we needed, and we expected most of our candidates to come from there. We did get applicants, but instead of the mid-career experts we were looking for, we only got much younger engineers. We loved their enthusiasm but were concerned about their ability to hold their ground with the Russians. We wondered why we did not see more of the mid-career cohort.

The interviews told the tale. Applicants reported that some of the flight directors were discouraging them, saying, "Anyone who takes this job can kiss their MOD career goodbye." This meant our applicants were taking a huge career risk just by talking to us. I was furious to see such hubris and partisanship putting NASA's mission at risk. When I described the problem to my wife, she was surprised at my reaction.

"What did you expect?" she said. "It is just another side of the arrogant and parochial attitude you saw when you were in their office."

She was right, of course. I went to see Holloway about taking it to MOD leadership.

Tommy shrugged and shook his head. For reasons I still did not understand, he was unwilling to engage MOD on any substantive issue.

"Leave it alone," was all he said.

There was plenty of skepticism about having the Russians in the station program, and we expected some naysayers, but this was more than individuals expressing personal opinions. This was technical leaders undermining the agency's formal mission, and I felt it warranted official action—even punishment. I continued to press Tommy to discuss it with the leaders of MOD and JSC, but as far as I could tell, he never did.

Having Gerstenmaier take the first tour of duty eventually brought the role the stature and credibility it needed. He made plenty of mistakes, but those were to be expected with a first effort. To make sure nobody missed the point, when he returned from his tour, I planted the seed that got him the biggest reward available—NASA's Stellar Award.

But that came later. We had to fill the next three ops lead roles with junior people willing to risk their careers for the success of the mission. I deeply admire and respect their courage and initiative. They took the risk and found their own ways to excel. They served us well.

While we were selecting and setting up our ops leads, the months were flying past. We continually made adjustments to the program—from changing how we managed food and clothing to developing training materials to tweaking our agreements with the Russians. The purpose of the Shuttle-Mir Phase 1 program had been to create the right conditions for success on the joint space station, now being called the International Space Station (ISS). Looking back, that was exactly what needed to happen, though, at the time, it seemed like chaos.

In the fall of 1995, STS-74 delivered a Russian-built docking module that allowed space shuttles to dock with Mir without having to change its configuration as they had for STS-71. Although the mission went superbly overall, the aftereffects of one small mistake would linger for years.

The best way to install the module was to first place it on top of the shuttle's docking system in the shuttle cargo bay, then dock it to the Mir using the mechanism on top of the module, which was compatible with the Mir's docking port. This procedure also served as a rehearsal for the way the first American piece of the ISS would be docked to the Russian segment a few years later.

The plan went well at first. Approaching the Mir, Shuttle Commander Ken Cameron maneuvered to dock the top of the module to the Mir. Although this looked difficult, with the face of the docking mechanism twenty-five feet over his head, the tools NASA had developed made it relatively easy. A camera in the center of the docking hatch looked through a reticle (sighting aid) at a docking target on the other hatch. By putting the shuttle in the right attitude and flying the reticle to the target, the docking itself was fairly simple.

All of this had been planned

Naming Missions

Space shuttle flights were officially identified by the sequence in which they were created. "STS" referred to the official name of the Space Shuttle program, Space Transportation System, and the number indicated its place in the sequence. For a variety of reasons, the actual launch sequence did not follow the flight number. In this book, I will refer to Phase 1 missions by their STS number and ISS missions by the designation of their mission element, such as 2A for the first American assembly flight.

and trained to NASA's standard of excellence, and after trying it myself in the shuttle simulator, I was very comfortable with the procedure. Then, as the shuttle approached Mir, one of the flight controllers came up with an idea to add a little more insurance. Since the surfaces where contact would occur were not directly visible to Ken, the flight controller proposed that the crew position the robot arm so that its camera could see the docking from the side. It was a perfectly logical suggestion, and the flight director quickly agreed. Had it been worked through the MMT, it probably would have been accepted by everyone. Unfortunately, the decision was made in real-time by the flight control team, and nobody thought to tell either the MMT or the Russians.

As the shuttle approached the Mir, the Russian flight controllers were unhappy to see that the shuttle looked different from what they had been told to expect. The arm was sticking out where it could potentially be a hazard to their solar arrays. They allowed the docking to continue but later complained loudly that the Space Shuttle program had changed the plan without discussing it with them. They were absolutely correct to complain; even though the docking was successful, the potential for a major problem had been real. We admitted the error and vowed not to let it recur, but the damage to the team relationship would resonate for years.

As if to highlight the challenges to come, many of my American colleagues could not understand why the Russian reaction was so strong. What they saw as a simple error in communication, the Russians saw as a lack of respect. Although I had seen this from my first meeting with the Russian designer in Crystal City, I had hoped that we had overcome it. In fact, it would continue to harm our relationship for years to come.

After STS-74 was undocked at the end of the mission, Tommy Holloway pulled Frank and me aside.

"Brewster Shaw just told me he is going to retire as the Space Shuttle program manager. I have been asked to take over," he said. "Frank, I am recommending you as my successor in Phase 1, and I suggest you make Jim your deputy, at least for now."

We were stunned. "When will this be announced?" asked Frank.

"Friday," said Tommy. "We can tell Ryumin at the Thursday telecon, but we have to wait to let Brewster tell the rest of the team."

This was exciting and sobering all at once. I had no idea whether Frank would keep me on or find someone else, but I hoped this would let us fix some issues. In my view, Tommy had been letting too many problems fester.

We had our work cut out for us, both because the pace of operations was increasing and because of the problems left unaddressed by Tommy's management style. Since he had made all the decisions himself, nobody else had a full understanding of his agreements. Even Frank, who had been his deputy for some time, did not know all the details.

After a couple of weeks as his acting deputy, Frank asked me to stay on permanently. I was proud to accept the role, but I suspected it said more about the dearth of volunteers for this controversial program than it did about me. The pace of work was such that Frank could not afford to wait for someone else to become interested.

As we dug into our new roles, we were distressed to find how weak the agreements were that Holloway had made with all the organizations that supported the program. He had assumed that each party would simply work out the details on their own, with few guidelines or reporting requirements. Everyone had done their best, but the free-wheeling approach meant many of the decisions were not very good. The lack of requirements, feedback, or coordination explained the poor support Norm had experienced throughout his mission. They also set the stage for more problems that would be encountered over the remaining missions.

Frank and I began to review the program as a whole and quickly saw a string of problems looming on the horizon. Many traced to how Phase 1 had grown from the single Shuttle-Mir mission to include four follow-on missions without much thought. The minimal agreements that might be acceptable for a single mission were not adequate for more. We soon encountered a constant stream of problems in areas we thought had already been resolved.

One of the most frustrating issues dealt with our crew's medical certification. The Russian doctors kept finding trivial issues to complain about, often threatening to disqualify our astronauts. NASA's doctors did their best to address each as it came up, but they were certain the issue was not medical but political. It was just another way for the Russians to demonstrate their absolute control. The string of nit-picking issues was maddening, but we patiently worked through each one, hoping to build the goodwill we would need in the future.

Other problems required that we create a new approach that was far beyond what had been planned. For example, NASA had agreed to supply food to our crews. Obviously, the menu had to be healthy and appealing, but personal taste, cultural sensitivity, and logistics were also important. Holloway had made an agreement with the life sciences team to provide the food for the crew, but he had not specified how much, what kind, or whether new foods could be prepared. Facing a long mission on a strange spaceship, some astronauts asked for new foods to add variety to their menu. The life sciences team was happy to oblige, but because they had no budget guidance, they spent nearly half a million dollars just for one man's requests. Once we discovered this, we put limits on future requests but were soon hit with other issues with this apparently simple detail.

For example, nobody anticipated that the American and Russian crewmembers would want to share each other's food. While this was undeniably a great way to build crew spirit, it created many new issues. The Russian doctors had different standards for crew diets and even what was considered acceptable food, including concerns about canned fruit adding tin to the diet. Although each issue was small, there were so many that we seemed to be caught in quicksand.

But the flight schedule was set, and the Space Shuttle program was keeping up, so we had no choice but to keep moving forward, even when our enthusiasm flagged.

12. SHANNON LUCID

The next crewmember to fly on Mir was Shannon Lucid, one of the first female astronauts in the US. She was a superb choice, with a PhD in biochemistry and four shuttle flights under her belt. Her enthusiasm and obvious love for spaceflight helped keep things moving when we needed it most.

Her program benefited enormously from Norm Thagard's experience, and many of the issues he'd faced were avoided for her. Alas, there were many new issues to be discovered, and our learning process accelerated into new dimensions as the NASA team became more involved in mission planning and execution.

Rick Nygren's team followed the traditional NASA approach to preparing Shannon's research program. This included the procedures she would follow and the training she would receive. What they failed to anticipate was the Russian demand to approve every nuance of the experiment and its procedures, including every change along the way. This might have been a modest issue had the experiments been mature, but they were not. Instead, the combination of new hardware, new procedures, and multiple Russian reviews caused the procedures to remain in a constant state of flux. Each cycle took a month or more, rapidly consuming our schedule reserves.

We also found it surprisingly difficult to create procedures whose meaning was exactly the same in both English and Russian, which was critical to gaining concurrence. A simple expression such as "turn on the system" could be misinterpreted if written as "close the power switch." Anticipating the challenges to come on the ISS, we began a major effort to define a common vocabulary and syntax for all joint procedures. It ultimately succeeded but was years in the making.

Shannon added her own element to the problem. While her final procedures were delayed, she was asked to train using draft procedures. Deeply invested in NASA's traditional approach, she refused to do so, arguing that training on draft procedures could plant incorrect habits in her head. When discussing this with Frank, I argued that this was why procedures were written down rather than memorized, but he deferred to her preference. We both had confidence in her ability to get the job done, but neither of us anticipated the ramifications of her choice. In the meantime, instead of training in specific procedures, she used her time to study the hardware and understand the research goals so that she could quickly apply the procedures when they arrived.

As a result, she did not see most of the procedures until she was on Mir. The final NASA approval for the mission, the Certification of Flight Readiness (COFR), documented that we had not met the normal standard for training. Although safe, some people objected that it could impact the quality of the research. They need not have worried; Shannon quickly understood the procedures and did a masterful job conducting the research. Although this initial trial had been painful and awkward, it was a first step toward developing better training processes for long-duration missions.

Her mission was also an important step in warming relations between our teams. Many Russians felt Norm's criticism had been unfair because they had treated him the same way they had treated other guest cosmonauts. We did our best to explain that Norm had not been criticizing the way he'd been treated but had been showing that what worked on short missions was not necessarily right for longer ones. We emphasized that he'd been reporting facts, not making value judgments. Some Russians accepted that, while others fumed.

Shannon's happy personality was the perfect antidote for the hurt feelings. The Russians made an extra effort to ensure her mission went smoothly. Her two cosmonaut crewmates, Yuri Onufrienko and Yuri Usachev, treated her like a welcome and valued member of their crew, and they got along beautifully.

Still, a few rough spots remained. When Frank and I arrived in Star City for her graduation ceremony at the conclusion of her training, Valery Ryumin took us into his office and closed the door.

Speaking through our interpreter, he said, "Shannon is very competent and well-liked by her crewmates, and we are confident she will have a good mission. But you should know that you are going to hear one criticism in the formal commissioning ceremony. Her language skill does not meet our usual standard for guest cosmonauts."

Frank looked concerned. "Is this a problem?"

"*Nyet, nyet*," Ryumin said through our interpreter. "Her crewmates have worked hard to ensure they can communicate, and they are confident they will be okay. Still, we must report that she does not meet the formal standard."

Frank glanced at me with a mixture of surprise and relief. Although we did not like hearing of the issue, we were pleased to see that Ryumin was trying to be a good partner. Once again, adversity was the engine that drove progress.

The commissioning ceremony was formal, with the presentation of the crew and

Learning to Trust

The Russian objection to the test we wanted to perform stemmed from concerns that the radiated signal could interfere with the performance of the Mir systems. This interference is called electromagnetic interference (EMI). My NASA colleagues believed it was merely an excuse to deny us something we wanted. However, through this conversation, I learned that Mir used older analog systems, which were far more sensitive to EMI than the digital systems commonly found on NASA. His concern was real, and I respected his responsibility.

their acceptance by senior Russian leadership. Each member stood and made a short speech about their commitment to the upcoming mission, and while Shannon's speech was rough, it met the requirement. Everyone seemed pleased.

In fact, we soon saw how popular Shannon was with the Russians. They celebrated her graduation with a party that featured dozens of homemade foods and desserts. We were relieved to see that she was headed for a good mission.

During this trip, I attempted to use our improving relationship to gain approval for other things we wanted to do on Mir. For example, because we wanted to use wireless communications between laptop computers on the ISS, an early version of the now ubiquitous Wi-Fi link, we asked the Russians to let us test the system on Mir. Their answer was a firm *"Nyet!"*

Rather than accept this rejection outright, I asked for a meeting with the Russian expert in electromagnetic interference to see if I could win his support for the test. We met in the Russian mission control center where he worked, discussing the idea through an interpreter. Although he was pleasant enough, he wasted no time telling me that he was opposed to the test.

Shortly after we began our discussion, our interpreter was called away to help with an urgent matter. Since I spoke little Russian and he spoke no English, we were helpless. After a few moments of embarrassed silence, he asked, *"Sprechen Sie Deutsch?"*

Because I had studied German in high school, my initial response was, *"Ja!"* Alas, the many years that had passed since high school and the technical focus of our conversation kept us from making any real progress.

But something important happened anyway. Even though our conversation was limited, we made a sincere effort to communicate. When the interpreter returned, he sat down with two different men than he had left just a few minutes before.

Working through the interpreter once again, I addressed my Russian colleague slowly and sincerely. "You look as though you are about my age. Do you have children?"

"*Da, dva.*" Yes, two, he said.

"So do I. Do you want them to grow up in the world we did, with the threat of nuclear war hanging over them?"

"*Nyet,*" he said, looking at me intently.

"Neither do I. We have been given the chance to do something truly profound—to build a partnership between our countries that can overcome the legacy of the Cold War. If we do this right, we can create a partnership and friendship that will secure the future for our children. That is my goal here today, not just getting your approval for a little experiment. Please work with me. Let's find a way to cooperate on this."

As he looked at me, I saw a change in his expression. He had heard my plea.

"Yes, I will work with you," he said through the interpreter. And he did.

My success with the engineer seemed to mark the beginning of a new, more productive phase of our program. Shannon likewise benefited from the great support she received from Bill Gerstenmaier. His familiarity with every aspect of her mission allowed him to anticipate her needs. Norm's team had wanted to do the same for him, but without the new agreements and additional help Bill enjoyed, they had been handicapped from the start.

Bill jumped into his new role as her ops lead with great enthusiasm. He not only knew the research plan, but he also knew the experiment hardware as well as she did. This knowledge allowed him to smooth out the rough spots before they became problems. A first-class engineer, Bill did his best to stay one step ahead of her needs. They made a great team.

Although I had known Bill for a few years, this was the first time we had worked together. The experience revealed a side of him that would come back to haunt me in the future. His total mastery of the engineering details was an enormous asset, but his tendency to hoard information for his own use quickly became a liability. When he was available, we had access to his vast knowledge, but when he was not, there was no substitute. He also tried to control everything personally. As the Americans in Moscow said, "When Bill leaves the building, the entire brain trust goes with him." This was a minor inconvenience for Mir, but it would become a bigger issue later.

There was one more surprise ahead of us with Shannon. The launch of the shuttle flight that was to return her was delayed twice— once when its solid rocket motors had to be replaced and again by a hurricane. The resulting delay extended her stay on Mir by six weeks. Since she had already completed her research program, she had little to occupy her time. Fortunately, her enthusiasm for spaceflight kept her spirits high, and she enjoyed her extended mission.

Once again, we had exactly the right people in place at the right time. Shannon's love for space, her easy relationship with her Russian colleagues, and Bill's hard work helped heal the rift that had arisen from Norm Thagard's initial foray. We looked forward to the next mission being even better.

But it was not to be.

13. JOHN BLAHA

When Shannon Lucid returned to Earth in September 1996, we were delighted to see that she was in great spirits. Her enthusiasm was genuine, and when asked in a post-landing news conference if she would be willing to go again, she said she would be ready in a month. Although her husband was not so sanguine, we were pleased with her positive attitude. Her smile showed that we had successfully addressed the worst of Norm's complaints.

Her replacement on Mir was John Blaha, a shuttle commander and former air force fighter pilot. Quiet, reserved, and somewhat ill at ease with the Russians, John seemed an odd choice for Mir. Although I barely knew him, I was concerned about his stay because serving as a guest cosmonaut on the Russian station was about as far from being a shuttle commander as a person could get. I worried that his low status in the Russian system would wear on him and lead to conflict. His friends told me not to worry, that John had a quiet boldness that would help him through any difficulties, and they were right.

Some of my concerns were eased when I learned that he had volunteered for the new program as soon as he'd heard about it. As a fighter pilot who'd served during the height of the Cold War, including combat in Vietnam, he saw the opportunity to work together with our former adversaries as an excellent way to help heal the wounds of that stressful time. Like most who had been a part of it, he was glad to see the Cold War end and to contribute to the ensuing peace.

John also looked forward to spending more time in space. He'd loved every minute of his space shuttle flights and had always dreaded having to come home. The idea of being able to spend months in space seemed like a dream come true. His original goal was living on Space Station Freedom, but his age and the delays the station program was encountering made it likely that flying on Mir was the only opportunity he would get for such an extended stay. He gladly accepted the challenges of living and working on a Russian station as the cost of achieving his dream.

There were certainly plenty of challenges for all of us. One of the most frustrating was the difficulty of making decisions stick. Often, the best we could get from our Russian colleagues was grudging agreement on individual events, not lasting changes in policy. This meant that many problems we thought had been resolved were actually postponed. They would often reappear in a slightly different form, sometimes repeatedly, and often when we least expected them. But while this was incredibly frustrating, when we did finally get an agreement, it was rewarding in its own way. We were pioneering on the frontier of international space cooperation, an extremely messy business.

One huge success was the operations team that Rick Nygren put together. He stepped up to the challenge that Tommy had given him and assembled a very capable group. By the time Blaha flew, he had a good team in both Houston and Russia. We expected much better support for John.

We also engaged more of the JSC team to work on innovations to make life easier for the crew and their families. We could not expect all our crewmembers to have the strength of Norm Thagard or the easygoing resilience of Shannon. We needed the program to take care of everyone we were going to send.

Shannon had already benefited from one of our best ideas: using a dedicated laptop computer to serve as a personal support system. It contained work-related things such as flight procedures, training materials, and reference documents but also things to brighten her day. These included personal items such as favorite movies, family pictures, home videos, and greetings that would pop up on special occasions like anniversaries, birthdays, and holidays. The idea was so obvious and worked so well that we regretted that it had not been available for Norm.

The ops lead concept had also proven itself with Shannon, and further refinements made it even better. Now that the Russians understood what we were doing, they became more accepting, sharing information earlier and sometimes even asking our preferences before taking action. Although there were still plenty of rough spots in our relationship, things were definitely improving.

Part of that improvement was because our formal relationship with Russia had changed. Norm's Shuttle-Mir mission had been a quid pro quo arrangement, meaning both parties had contributed without receiving any funds from the other. While easy to implement, this arrangement meant that we'd had little leverage. The agreement for Phase 1 made us paying customers, allowing us to demand that the Russians fulfill their contractual obligations. Ryumin understood this in principle, but turning that understanding into concrete actions was difficult, requiring hard negotiations to secure even our minimum needs.

Jim Nise, a gregarious former naval flight officer, had the unenviable job of ensuring the Russians met their contractual responsibilities. This was an enormous leap for the former Soviets, but Jim brought great energy and good humor to the task. His amiable but firm stance usually did the job.

Meanwhile, I continued to look for ways to create a more cooperative effort beyond the strict limits of our contract. One of my goals was to show Ryumin that NASA brought new capabilities to their work that could carry forward into the ISS program. My first initiative began right after Norm's mission.

Norm had emphasized how his stay had suffered from the lack of communication. The Russian reliance on ground sites within their national borders meant that Mir could only talk to the ground when it was within line of sight to their antennas, with the average communications pass lasting under ten minutes. Worse, over the course of a day, there was a period of about ten hours with no communication at all. This was a major inconvenience that could become critical in an emergency.

Seeking a way to improve the situation, I investigated whether NASA ground sites could augment the Russian ones. To build cooperation, my plan was to make them available to the Russians as a common resource, hoping they would view us as valued partners who could contribute to their operations. While I knew that such a change in attitude would take years, the sooner we got started, the better.

As soon as I got management approval, NASA's can-do spirit kicked in, and communications experts looked for the fastest and least expensive way to put the capability in place. Three sites agreed to put new antennas on existing mounts and use commercially available amateur radios to connect with Mir. The sites would then automatically relay the communications to the Russian mission control center. Many employees at these tracking sites were amateur radio enthusiasts who contributed their time and effort at no cost. I was delighted with their enthusiastic support.

Although I expected the Russians to be pleased by the effort, we waited to tell Ryumin about the effort until we thought it was ready. When we explained that we had done it to support both our crews and emphasized that it was at no cost to them, Ryumin accepted the offer but was clearly unimpressed. We presumed that he did not want to show any sign of weakness or need for our help but hoped it would lead to more collaboration in time.

The early attempts to use the US sites did not go well. At first, the quality of the audio coming through the US stations was poor. Our team was frustrated and embarrassed because they thought we were duplicating the Russian capability exactly. They tweaked the radios, improved the pointing of the antennas, and looked for sources of interference, all without much benefit. Yet even though the quality was poor, the Russians could see us trying to help, which was important.

Eventually, Mark Severance, who'd spent many months supporting our operations in Moscow, figured out the problem. He redid the analysis of the communications system from one end to the other—what engineers call a link budget—and uncovered a missed detail. Once it was corrected, the American communications worked fine and became a routine asset for the program.

Although John was relatively well-prepared for his stay on Mir, his mission almost got off on the wrong foot. Upon his arrival, while STS-79 was still docked to Mir, he checked his experiment hardware and discovered that one of his primary experiments, the biotechnology system experiment—designed to grow cells in microgravity—was not operating correctly. The problem seemed severe enough that there was talk of returning it on the space shuttle if it could not be repaired. Losing this experiment before the mission even began would have been an enormous blow to his morale, so the team's efforts to restore it reached a fevered pitch. As the time for undocking approached, yet another fix was suggested, and John threw himself into the effort. He worked on the experiment right up to the time the shuttle undocked, even forgoing the farewell ceremony. His effort was successful, and the experiment was saved.

Once that excitement was over, John's mission settled into a fairly routine pace. But even a routine mission on a foreign space station presents many opportunities for trouble. While Norm's mission had been a pioneering effort with numerous small problems and Shannon's mission had given us a chance to fix some of the obvious issues, John's mission warned us not to become complacent.

> **Medical Privacy**
> This book follows NASA's strict policies prohibiting the release of private medical information, including psychological states. I am free to mention John's challenges only because he has openly discussed them with the media and the public.

His experience reminded us that every person is different and that every situation can present unique challenges.

Perhaps the biggest surprise for us was how his isolation from family and colleagues led to loneliness and depression. All the Americans who flew on Mir felt this isolation, but John's experience was exacerbated by both his personality and the fact that he was not flying with the cosmonauts with whom he had trained. A few weeks before their launch on a Soyuz, his commander, Gennady Manakov, had been grounded due to a heart problem. According to Russian protocol, when one crewmember was unable to participate, the entire crew was replaced with its backup. The cosmonauts who took their place, Valeri Korzun and Alexander Kaleri, were highly competent and accepted John easily, but it was still a major disruption. John had barely met this crew and had only a few days before their launch to get to know them, leaving little time to establish a personal relationship. What was already going to be a significant personal challenge was now even more so.

Fate had one more surprise, which made John's time even more challenging. His ops lead, Isaac "Cassi" Moore, an intensely focused MOD engineer, had been eager to support the program at the Russian mission control center. However, he found that his extended time away from home put too much strain on his young family. By the time John arrived at Mir, the months Cassi had already spent away had taken their toll, and he felt the need to return home. We were surprised and disappointed but respected Cassi's decision to prioritize his family. He continued his support from Houston, spending many hours on the phone with his colleagues in Russia, and we added another item to our watch list—ensuring we could replace critical people if needed.

For the most part, NASA has done well in selecting strong and capable people as astronauts, and John lived up to this standard. Together with his cosmonaut colleagues, John acted as a true professional, overcoming many challenges to complete a successful mission. He worked hard to integrate with his new crew, conscientiously contributing to their success as well as his own.

Despite the challenges, John thoroughly enjoyed the experience of living in space, especially his role in supporting two spacewalks the Russians conducted to upgrade and repair Mir's systems. The sight of his cosmonaut colleagues working outside the station, with the beautiful Earth in the background, made a powerful impression—one he would share many times after returning home. He was also diligent in his research and made good progress despite the limited support that continued to plague the Phase 1 program.

While John's experiences taught us new lessons, and there were still plenty of rough edges, we were beginning to believe that we had the basics of working on Mir under control.

The next mission would dispel that illusion.

14. JERRY LINENGER

After John Blaha's uneventful mission and return to Earth on STS-81 in January of 1997, we began to feel as though we had both a sound template for our work and a sustainable working relationship with the Russians. Although we had a long way to go, at least we were making progress.

Blaha was replaced by Jerry Linenger, a friendly and generally easygoing medical doctor who was something of an enigma to most of us. Each of the other crewmembers had flown multiple missions and was well-known to the community. Frank Culbertson had personal relationships with most of the other astronauts going to Mir and could give us insight into what they were thinking and how they were likely to react. But Jerry was not so well-known, even to Frank, and we would soon find it difficult to anticipate his thoughts and actions during the mission.

Throughout the Shuttle-Mir Phase 1 program, the Russian program suffered many problems, most of which could be traced to the economic stresses that followed the fall of the Soviet Union in 1991. Their hardware was aging, suppliers were becoming difficult, and skilled engineers were harder to retain as the Russian economy worsened. We were concerned about the potential problems this could cause, but the proud Russians refused to discuss the issues or the possibility of help that we could offer.

Because one of our explicit goals was to learn from the Russian experience of operating space stations, we were frustrated that they

132

shared so little about their program and how Mir's systems worked. Some American engineers working on ISS were learning some of the basics about the Russian systems that would fly on their vehicle, but the Russians we dealt with were very tight-lipped about Mir's systems and their performance. We thought this silly since in many cases, the systems were the same. The relationship that Bill Gerstenmaier had built with the Russians had started to open the dialogue, but it had been personal. When he left, it had to be recreated anew by each of his successors, which was difficult and time consuming.

We were particularly distressed by the way the Russian operational leadership continued to hold us at arm's length. They utterly refused to let us be a part of the mission planning or management even as they wreaked havoc on our research program. In our weekly telecons, Frank would often ask for the opportunity to participate in the planning so that we could be more effective, but Ryumin would respond that we had no role in the spaceship's operation.

Many of our American colleagues thought the reason for this was that the Russians were embarrassed by their primitive facilities and systems. This reflected the common US view that NASA's way was best, but I disagreed. My interactions with the Russians convinced me that they actually felt superior to NASA and did not want us to take advantage of their expertise. Having been first to put a man in space, many Russians believed that their lead remained absolute. They also took special pride in the fact that their accomplishments came without the bountiful money and fancy technology we enjoyed. Regardless of the cause, they blocked most attempts we made to increase cooperation with them.

I was distressed to see pride threatening the mission. They should have been sharing their knowledge with us in order to forge the operations team we would need for ISS, but they did not.

In truth, many Americans were equally disinclined to cooperate. This parochialism, while far from universal, was sufficiently pervasive enough on both sides of the ocean that we worried about it creating new risks for the coming International Space Station. Frank and I were glad for the opportunity to work through such issues on Mir rather than on ISS. But we both worried that it might not be sufficient.

On February 23, 1997, after Jerry had been on board for about a month, a new Russian crew arrived at the Mir. Their Soyuz would stay there throughout their mission while the old crew would return in the one that had brought them. By keeping a fresh Soyuz always on the station, the crew could return at any time in an emergency. This overlap between the crews also gave the old crew the opportunity to bring the new crew up to date on the Mir's status.

This handover process also meant that there were two crews on board for about a week, which put extra demand on the life support systems. These systems were designed to support a normal crew of three, so larger crews required that they use backup systems to meet the extra demand.

Mir's oxygen for breathing was normally supplied by a system called Elektron that used electricity to break water into oxygen and hydrogen. The oxygen was released into the cabin and the hydrogen was vented overboard. It worked fine but required quite a lot of power and cooling, which was not always available. Its backup system used simple chemical cartridges to produce oxygen by chemical reactions that also produced heat. Each cartridge made enough oxygen to sustain two people for one day while using only a tiny amount of power for a fan.

To use the backup oxygen source, the crew simply put the cartridge into a wall-mounted holder and activated a striker to ignite it. Once the reaction started, a fan blew air through the unit to cool it and spread the oxygen around. It was an elegant solution to the backup oxygen requirement.

However, on this occasion, instead of the cartridge producing the expected rush of friendly, warm oxygen, a bright jet of flame burst forth into the aisle. This flame burned fiercely, igniting trash bags and other loose items and blocking access to one of the Soyuz capsules.

Fire in the confined spaces of a spacecraft can be catastrophic. Not only can it spread to other flammable materials, but it can also damage critical systems and structures. If it impinges on the wall, it can soften the aluminum enough for it to rupture from the air pressure within. Fire also creates toxic gases and airborne debris that contaminate the atmosphere and reduce visibility. The result is a nightmare scenario in a small, sealed volume, with no opportunity to escape or even vent the fumes.

The crew had been well-trained, and as soon as the alarm sounded, they all went to work. While the cosmonauts did their best to control the fire, Jerry raced to the other modules to retrieve fire extinguishers and breathing apparatus for them. His support task, which should have been simple and quick, turned into a maddening, life-threatening ordeal.

The fire extinguishers were held to the wall with quick-release mounts similar to those found in public buildings around the world. However, when Jerry tried to free these in the new *Spektr* module, he found that the bolts installed to keep them from vibrating loose during launch were still in place. These bolts should have been removed during the module's initial activation, but they had not been. Instead of simply grabbing them, Jerry had to find tools to remove the bolts before they could be used to fight the fire.

Backup Systems

Spacecraft almost always have backup systems for critical functions to ensure safety following a failure. In some cases, the backups are of an entirely different design. Both Mir and ISS use primary systems to provide oxygen and remove carbon dioxide that can operate for long periods without consumables but which require significant amounts of power. Both spacecraft also include backup systems that can be operated for short periods with far less power. This ensures that critical life support functions can be provided even when electrical power is limited.

The extinguishers were designed to fight a conventional fire fed by combustible materials. They worked by cooling the fire's chemical reaction and cutting off its supply of oxygen. However, this fire was being fed by the heat and oxygen from the cartridge's internal chemical reaction, making the extinguishers almost worthless. The best the crew could do was keep the fire from spreading to other flammable materials as burning bits of metal flew out of the cartridge. The ordeal continued for nearly ten agonizing minutes until the chemicals in the cartridge were consumed and the flow of oxygen stopped. Finally, after what seemed like an eternity, the fire went out.

The next challenge was simply to stay alive in the highly contaminated atmosphere, and for that, the crew had to depend on a small number of emergency breathing masks that used a chemical to provide fresh oxygen. Each mask would provide oxygen for about an hour, but there were only a few of them, as they were only intended to respond to a modest and unlikely event. However, this fire produced much more contamination than anyone had anticipated. The entire station was filled with smoke and steam, requiring hours of filtering before it would be safe to breathe normally. When the oxygen-producing masks were exhausted, the crew was forced to rest quietly while breathing through simple filtration masks that used activated charcoal to absorb some of the contaminants.

The crew also had to decide whether to abandon the spaceship— an agonizing decision that might end the entire Mir program. Their decision was complicated by the fact that they were in one of those extended periods with no communications available through Russian ground sites, and even the next American site was hours in the future. They decided to stay despite having no way to ensure that there were no poisonous gases from the fire. It was a very courageous act.

Fortunately, as a medical doctor, Jerry was able to monitor their condition and treat their injuries. Once communications were restored, he advised mission control on their health. Everyone knew that their exposure to the thick smoke could produce potentially serious respiratory effects, and we were relieved that Jerry was there to watch for them.

The first opportunity for the crew to talk to the ground came through the NASA Wallops Flight Facility. A technician at Wallops heard a garbled but animated message coming from Mir. The audio was automatically relayed to Moscow, where, despite its poor quality, Russian mission controllers made out the word "fire." Although they were unable to respond to Mir at the time, this alerted them to the emergency. They quickly assembled the experts needed for whatever might come next, including search and rescue teams ready to find the crew if they had to evacuate.

NASA's first hint of a problem came in the morning when Jerry's ops lead, Tony Sang, arrived at the Russian mission control center and overheard a conversation between controllers. Although his Russian language skills were still limited, he recognized the Russian word for "fire" and immediately went looking for the flight director to learn more. Although they did not have much to share, Tony immediately called Frank Culbertson at his home in Houston. It was about five o'clock in the morning when Frank called me to join him in his office so we could prepare to respond.

When I got there, Frank was on the phone with George Abbey, the Johnson Space Center director. Frank's expression communicated more about the seriousness of the event than words ever could. I heard him updating George, and when he got off the phone, we built a plan for the day.

"I'm going to have my hands full talking to the press and headquarters," Frank said quietly. "I need you to gather all the information you can on the oxygen generators, the fire extinguishers, and all the rest of the system."

"Will do. I've already sent emails to MOD and the engineering contacts to meet in my office. Have we heard anything from the doctors?"

"Rick Nygren has a call in to them. Nothing yet, except we are lucky to have Jerry there. He is monitoring the others."

The normal workday started early at JSC, so I soon started talking with our experts on Russian life support systems. People quickly gathered the information they had and came to my office. By eight o'clock in the morning, the room was filled to overflowing with MOD and engineering directorate colleagues who were experts in life support and emergency systems. After I filled them in on what we knew about the night's events, I went around the room asking for input. Everyone shared all that they knew, but it was painfully inadequate. Compared to the absolute mastery these people had of NASA systems, they were embarrassed to have only general descriptions of the Russian ones. Our joint sense of powerlessness was palpable.

Most NASA people, especially those in MOD, feel utterly responsible for the safety of the crew and the success of the mission. This urgent responsibility, so aptly summarized in Gene Kranz's "Tough and Competent" motto, was being frustrated by our inability to get the information they needed to understand this crisis. They were mortified that they could not answer the questions I was asking. As we stammered and guessed, I resolved to somehow use this event to overcome the barrier with Russia so we could be full partners in any future problems.

While I worked on our response, Frank spent the day briefing NASA management, conducting interviews, answering questions from the press, and trying to calm congressional staffers. Because this was our first major emergency, our answers were accepted without too much challenge.

As in any chaotic emergency, some bad information found its way into the reports. Someone in Russia had gotten the impression that the fire had only lasted about ninety seconds, so that was what Ryumin told us and what we told everyone else. This suggested that the crew had been successful in extinguishing the fire fairly quickly, rather than the ten minutes of terror they had actually experienced.

A few days later, Jerry held a press conference from Mir. When he heard the media had been told the fire was just a short event, he was furious. Although he controlled himself in front of the press, he was extremely upset because, to him, this smelled of a cover-up. We tried to assure him that it was a simple misunderstanding, but the issue would not die. Ryumin and other Russian leaders insisted for weeks that the ninety-second figure was correct and that the fire had been extinguished successfully. It was not until an investigation proved this was impossible that they reluctantly accepted the crew's version.

Because our communications with Jerry were so limited, we struggled to reconcile the different versions with him. The problem worsened when he, like Blaha, decided not to talk to us. He shared very little with his ops leads and virtually nothing with Frank and me. Had he been more communicative, we would have understood much more about the fire and his situation in general. More importantly, we would have been able to maintain his trust. Instead, it was months before we had a good conversation with him, during which he assumed we knew what was going on. He resented what he perceived as our lack of support for him, which colored the rest of his time on Mir.

We insisted on a full joint investigation of the fire. This serious mishap had put the crew's lives in jeopardy, and the future of the program was in doubt. At first, the Russians refused to let us participate, but Frank made it clear to Ryumin that we could not continue flying people to Mir unless we fully understood what had happened. Since they desperately needed the funding we provided, they eventually agreed.

The investigation showed that the fire had most likely been caused by contamination in one canister and was unlikely to recur. However, it also found procedural weaknesses that might have contributed to the problem, including crews storing the cartridges without their protective packaging. Additional steps were taken, including the creation of procedures to better protect the canisters to reduce the chances of further problems.

The fire was just the first of Jerry's challenges. His time on Mir coincided with some of the lowest funding the Russian program would receive from their government, and the station was showing the effects of age and poor support. Over the next few months, his Russian crew was constantly making repairs to the life support systems, which broke down with increasing frequency. Many of these repairs were jury-rigged to keep the systems functioning until new components could be brought up on the shuttle or one of their Progress resupply vehicles.

At the same time, the cooling system developed pinhole leaks in its tubing, allowing coolant to get into the atmosphere. This coolant contained ethylene glycol as an antifreeze, similar to that used in automobiles, and we were concerned about the health effects of breathing it. To our surprise, the unanimous conclusion of the doctors was that, while ethylene glycol is highly toxic if ingested, it would become an intolerable irritant to the eyes long before it reached toxic levels through inhalation. If the crew could tolerate the irritation, they were safe from serious health effects.

This was relatively good news, but we wanted direct feedback from Jerry. This we never got, as he continued to keep us at arm's length. He reported that the glycol effects were tolerable, but we did not learn until he returned that he thought they were only barely so. This was another case of miscommunication. We were looking for his expert opinion, while he assumed we were seeking his agreement to endure some discomfort.

The problem with air quality was annoying in itself, but it also had an unanticipated effect on crew morale. Because a higher heat load would make the cooling system work harder and worsen the leaks, the Russians decided to limit the crew's exercise periods. Nobody liked this, as exercise is important for both physical and emotional health, but Jerry objected strongly. While we knew he was fanatical about his physical condition and wanted to walk off the shuttle at the end of the mission, we could not advocate special treatment for him over his crewmates.

Jerry went around us and complained to his crew, who then relayed his concerns to Russian management. As a result, the Russians further decreased the cosmonauts' exercise time to allow Jerry to exercise daily. His crewmates were deeply offended, further undermining the already-strained working relationships on board.

For excellent technical reasons, we were less concerned about the reliability of the life support systems. In the worst case, if the backups and consumables were exhausted, the crew would have to return home on the Soyuz. This would be bad for Mir, but it did not pose a threat to the crew's safety. They would have days of warning, allowing plenty of time for a safe retreat.

Some critics of the program exaggerated the nature of the problems and their urgency. They referred to the Soyuz as a "lifeboat," implying it was a last-ditch rescue vehicle for the crew in an emergency. This was blatantly misleading and created unnecessary pressure on an already-stressed relationship. Whenever possible, I emphasized that the Soyuz was the vehicle the Russians used to travel to and from Mir, not a lifeboat. To me, it was analogous to a car used to reach a mountain cabin. If you ran out of food in the cabin or mosquitoes became unbearable, you took the car and went home. It was not an emergency, just an inconvenience.

Of course, it was not quite that simple. What would be an inconvenience to us could be a program-ending crisis for the Russians. If they abandoned Mir, they might not be able to return to repair it, bringing an ignoble end to their premier space station program. We worried that the pressure on management might cause them to take greater risks than we thought prudent in order to protect their program. We were constantly on the lookout for such risks but could not see any clear evidence of them. We investigated every claim brought to us and found none had merit based on what we could see and understand.

The inflammatory chatter severely complicated our efforts to make good decisions because it energized our critics, who were universally less informed than we were. Congress was concerned, but they generally accepted our well-argued judgments because they knew they were not competent to comment on the issues. But some senior NASA retirees, including Apollo astronauts, ignorantly repeated the inflammatory and inaccurate criticisms. Some even demanded an end to the program. This struck us as hypocrisy, as these same people had taken far greater risks with both Apollo and the early Space Shuttle program. We spent a lot of time addressing their arguments with our management and the press.

The safety of the Mir itself was not our problem, but the safety of the crew was. As long as the crew had the option to safely return on the Soyuz, we felt it prudent to continue. The urgency we felt was the need to gain from this painful experience before putting the ISS at risk. The new space station represented a far greater national and international investment, and we felt strongly that the joint operating experience was worth the small risk. As long as our crews were willing to go, we wanted to continue.

15. STS-84

Jerry Linenger was scheduled to return to Earth on STS-84 in May 1997, which would also deliver his replacement, Mike Foale, to the Mir. Although we saw it as just another flight, it sparked a great debate on the health of Mir. Both Mir and the Russian system that supported it were under great stress, and it was difficult to determine the cumulative effects of that stress. We were concerned about potential hardware failures and human decision-making, but we still saw a valuable learning experience for the ISS.

Mir had a history of long-term problems, such as small power margins, which worsened as the solar arrays aged, and reduced cooling in the main living quarters, known as the base block. The power problem was resolved with the new solar arrays on the *Spektr* module, but for some reason, the cooling issue was never fixed. This caused uncomfortably high air temperatures and humidity in that module.

Our crews reported that the temperature was irritating but bearable, yet the continuing stress in the Russian program led us to pay closer attention to the cumulative effects on the spacecraft itself. Although the Russians rarely talked about it, we learned that a large amount of water was condensing on the interior walls of the module, and the problem seemed to be worsening.

The temperature of the outer shell was controlled by a water/ethylene glycol mixture that flowed through tubes welded to the shell's inner surface. These thin aluminum tubes corroded when in contact with water and certain metals. This corrosion created a growing number of pinholes, allowing coolant to seep into the cabin air. The leakage increased humidity, causing more condensation and an unpleasant smell.

The smell of the coolant had been noticeable during Shannon Lucid's stay a year earlier, but it was getting worse. With no instrumentation available to gauge the severity of the problem, our assessment relied on the crew's reports on the intensity of the odor, the level of irritation it caused, and the amount of coolant needed to refill the system. Based on this information, we concluded that the issue was concerning but not yet severe.

The Russians were constantly repairing leaks using an epoxy paste to seal the tubes. The technique seemed to work but was slow and tedious, as each hole had to be located and treated individually. Its biggest disadvantage was that it required turning off the coolant loop long enough for the epoxy to cure. This exacerbated the already-high temperature and humidity, making the crew miserable, and the increased condensation promoted new leaks.

Ryumin finally admitted that many Mir systems needed replacement. Some Americans interpreted this as a sign of poor design, but it actually reflected a conscious decision by the designers. While NASA invested large amounts of time and money to develop long-lived systems, the Russian approach was to accept cheaper systems and replace them as needed. This worked fine as long as replacements were available, but it became a problem when they were unavailable or unaffordable.

The most urgent problems were with their life support systems. Oxygen came from the Elektron system, and carbon dioxide was removed by the Vozdukh system. Both were complex, consumed significant electrical power, and had a modest lifespan. Now that they were being pushed beyond their normal lifetimes, they broke down frequently.

Ryumin admitted the units were well beyond their planned replacement ages because budget constraints had forced them to stretch the old units as long as possible. Daily reports showed the Russian crews constantly swapping components and making temporary fixes in an effort to gain every bit of life they could.

While this was tedious for the Russians, we learned valuable lessons from watching them finesse their systems for maximum longevity. I was delighted by the priceless education we were gaining on the challenges of operating a long-duration spacecraft, and I assumed my NASA colleagues were taking full advantage of the opportunity.

Tracking this activity became a central focus of our work so we could intelligently oversee the safety and progress of our missions. Most of what we learned was both logical and reassuring. With two identical systems for each function, the crew could keep one system operational by cannibalizing parts from the other. When the same parts failed on both systems, they could sometimes recover enough working components to get one working. At other times, they were out of luck, but they had enough cartridges for the backup systems to last for several weeks. While this was manageable for short outages, it could be a problem if there was a long wait for spares to be launched. Our shuttle became a valuable alternative for getting supplies to Mir because it arrived during the interval between Progress flights.

Initially, the Russians tried to downplay their problems, insisting they would be fine on their own. However, the constant concern over the next failure taxed everyone's patience, impacting our research program and threatening the future of the Shuttle-Mir program. The effect on our research was often severe, so we told Ryumin that he needed to get Mir healthy to keep the support of our national leaders.

Those leaders were starting to pay attention. After the fire, Mir's condition became a topic of interest to many, from colleagues to members of the media and even Congress. As the May 1997 STS-84 launch date approached, we were flooded with requests for information about Mir's status and plans for restoring it to full functionality. Fred Gregory, a former astronaut and head of NASA's safety organization at headquarters, was particularly attentive. His job was to protect the safety of US crews, and while he and I agreed on the goal, we struggled to find a common approach to achieving it.

Fred and his safety engineers were used to NASA's "belt and suspenders" approach and chafed at the Russian approach of "good enough." They were also used to having unlimited access to information, often asking NASA programs to conduct additional tests to address their concerns. However, this was not the Russian way, and we had virtually no leverage to obtain more data, let alone conduct additional tests.

Fred and his team admitted that the system designs were satisfactory, but they did not like the Russian approach to maintaining them. Fred's headquarters organization had not established a good relationship with its Russian counterparts, leaving the Phase 1 team in Houston responsible for clarifying issues. This put us in a difficult position, especially when we thought the Russian approach was reasonable.

This awkward relationship suddenly became a major problem when Fred publicly stated that he would only support continued operations on Mir if the Russians met his specific requirements. What he wanted was logical from his point of view but completely unacceptable to the Russians. His requirements did not account for Russian experience and design philosophy, nor were they part of our intergovernmental agreement. Frank and I sympathized with his position, but we found the Russian approach acceptable and lacked the leverage to demand changes. The Russians knew Fred's position but ignored it, continuing to manage their program by their own standards. We had no choice but to comply.

Although I wished he had taken a different approach, I appreciated Fred's interest in the situation. While I was okay with the Russian approach to their own station, I was certain it would not be satisfactory for the larger ISS program. NASA wanted to manage ISS with the lowest practical risk, which was not what we were seeing on Mir. I had no idea who would be the manager responsible for ISS operations, but I knew that our actions could affect how hard the job would be. I hoped that Fred's public comments would alert the Russians to what would be needed in the future. Once again, Phase 1 was a crucial step in the transition to the new program.

While the Russians did not want our help with their planning, they did want our help to bring up their spares on the same shuttle that delivered our crews. I shamelessly used this as leverage, forcing Ryumin to inform us as soon as potential needs arose. This allowed us to show critics that they were not ignoring problems but had work underway to fix them. We flew a number of critical spares on STS-84, including an entire replacement Elektron, which helped quiet the critics and reassure our other partners.

Space Shuttle program management, however, fought our requests to add these Russian logistics because the appeals came close to launch time and violated the program's goal of finalizing the cargo manifest months before launch. The program's timeline had worked fine for previous missions that had been standalone and had rarely been impacted by events outside of the mission itself. They'd had plenty of time to plan every detail. But supporting a space station meant that NASA, including the Space Shuttle program, needed to be able to respond rapidly to unexpected failures on orbit.

Frank and I were convinced that this was the most important lesson the Space Shuttle program could learn from Phase 1. Operating a space station required a logistics system that could respond to unplanned events as quickly as possible. Shuttle management argued that late changes increased the risk of mistakes, such as flying the wrong item or damaging a piece of hardware. We argued that such rapid support was now their job, and they needed to get good at it. My planning for maintenance on Space Station Freedom had predicted the need for a rapid response, and this real-world experience proved it.

The Space Shuttle program personnel's reluctance to accept our requests for STS-84 might have prevailed without the management attention that came from the fire and other problems. Although I was confident that George Abbey would have helped us if we had asked, Fred Gregory settled the issue by declaring that the spares were now a NASA Headquarters requirement.

The Space Shuttle program reluctantly complied, and we hoped that we were making progress in changing their attitude toward the issue. We did not realize that the program saw this as a one-time exception, not a permanent change. They had no intention of modifying their processes to suit another program.

There were those who did their best to help. The most important of these was Spacehab, a commercial contractor that provided and operated the cargo module used for most of the Mir flights. Their entire team, from their president, Mike Kearney, to the technicians who stowed the cargo, bent over backward to accommodate the late changes at little to no additional cost. This often required them to develop new processes, including loading large cargo into their module after the shuttle was on the launchpad. They always did whatever we asked, sometimes without even waiting for our request. Kearney viewed this as part of learning to support a station, and he was glad for the opportunity. He often absorbed the cost of new processes as an investment in their future.

Two weeks prior to the launch of STS-84, we flew to Florida for the flight readiness review (FRR) at the Kennedy Space Center. Valery Ryumin was once again the senior Russian present at the all-day meeting, joined by our senior interpreter, Vladimir Fishel, a mutual friend and confidant.

Adding to Ryumin's interest in this launch was the fact that his wife, Yelena Kondakova, was flying as a member of the crew. She had spent much of the last year training in Houston for her second flight to the Mir and her first on the space shuttle. Ryumin had attended all our FRRs for shuttle flights to Mir, but this was the first time he was seriously interested in the health of the space shuttle itself. I could see he was paying close attention to every aspect of the daylong meeting.

Frank and I had split roles and responsibilities for the FRR. He was a member of the review board and sat at the table with the other senior managers while I presented the status of the program and our readiness to proceed. Although one might expect the Russians to present the health of their own vehicle, they refused to answer to an American management review, so the task of reporting the status of Mir and its systems fell to me. This was fine with us because we thought that Ryumin's brusque manner would not have been well-received by the board.

Because of the spring's events, we expected some extra interest from the FRR board, and I took extra time to describe the status of each system in some detail. To my surprise, the presentation had the full attention of the board, and a few of its members went so far as to aggressively challenge the Mir's readiness. Instead of providing only a boring twenty-minute summary, I spent more than an hour at the podium answering questions for which I was just barely prepared. While I was confident that the mission was appropriate and the condition of the Mir was safe, I had not anticipated such detailed questions from an aggressive audience. It was a challenging briefing, and I was glad when it ended.

My remarks concluded shortly after noon, and the board adjourned for lunch. As I was leaving the podium, Ryumin and Fischel approached. Ryumin grabbed my hand with his two gigantic mitts, looked me hard in the eyes, and spoke in Russian. Vladimir quickly translated: "You fought like a lion!" Valery shook my hand strongly before turning to leave, and Vladimir leaned over to quietly say, "That is the best compliment anyone will *ever* get from Valery Ryumin!"

After lunch, the FRR resumed with a discussion of anomalies that had been observed on the last launch. One had resulted in a dent in a cover used to retain debris from the separation of the shuttle from its external tank. The conversation was highly technical and went on for about thirty minutes, during which Vladimir did his best to translate the conversation for Ryumin. I watched to see if he was bothered by the conversation and could see no signs that he was. Even so, at the next break, he sought me out to ask some questions.

Stepping outside, I was glad for the opportunity to escape the oppressive air conditioning of the Operations and Checkout Building. Ryumin used the opportunity to enjoy a cigarette.

His questions addressed the anomaly that had generated so much conversation. He wanted to know if it threatened the safety of the mission. Even though Ryumin was an imposing man, having served both as a tank commander in the Soviet army and as a cosmonaut on some very demanding missions, his face and manner had subtly changed to show concern for the safety of his wife. He listened closely and asked many questions as I patiently described the system and the anomaly. I gave him the facts and was as thorough as I wished his people would be with us. He appreciated my openness and candor, and when he asked for my opinion on whether he should be concerned, I clearly and confidently told him no. His relief was obvious, and he once again shook my hand as we returned to the meeting.

Following the FRR, we and the management team returned to Houston to make our final preparations for the mission. Frank and I then returned to Florida a few days later to meet with the Russians who would attend the launch. The neutral venue and emotional energy of a launch, producing a bit more flexibility among the parties, made this the most productive time for a joint meeting. It was also an opportunity to socialize and get to know our new partners more intimately.

Two days before the launch, we had a dinner event at a facility known as the Beach House, a former private residence on the beach near the launchpads. Astronauts had spent time with their families there immediately before launches until it was converted into a conference facility for official events. That evening, we were visiting with our Russian and American colleagues when I noticed an older woman enter the room. Somehow, she struck me as familiar, and I blurted out, "She looks like Valentina Tereshkova!" One of my Russian colleagues said,

"*Da.* She is here to celebrate the launch of another Russian woman to space. She will be with Valery Ryumin when the shuttle launches. Valentina, come meet Jim!"

Tereshkova was a world-historical figure—the first woman to fly in space. During the evening, she was modest and polite to all, like someone's grandmother who had come as a guest rather than one of the most legendary figures in the room. When I briefly spoke to her, she was very gracious.

Two days later, we prepared for the launch. As usual, I escorted the Russians to the viewing site, where they no longer felt the need to stick together. They wandered individually through the crowd, taking in the sights. As the shuttle lifted off, I looked up into the viewing stand and saw Ryumin and Tereshkova huddled together with Ryumin's daughter.

The mission went very well. No significant problems presented themselves, and Mike Foale was delivered for his adventure on the Mir—an adventure that would change our world forever.

16. LOST OPPORTUNITIES

The end of Jerry Linenger's mission left me with mixed emotions. While I was glad to have him back after a successful mission, it also left me with an odd sense of foreboding. Between the hardware problems that had haunted his mission and his failure to communicate, I felt that we were somehow headed in the wrong direction.

His return also brought us a great deal of new information about his mission. It was normal to learn a lot in a debriefing, but his took on a whole new level of importance because so much of this program was new, communications were so poor, and Jerry and the Russians had shared so little. Timely and accurate crew inputs were vital to our success, yet once again, we'd been left in the dark.

Our already-limited communications with Mir had worsened when Linenger repeated Blaha's behavior of cutting off noncritical communications. He wrote emails to his wife, often in the form of "letters" to his son, an infant at the time, and chatted with his flight surgeon and ops lead, but he sent very little useful information to the rest of us. Apparently, he assumed we knew what was going on as if it were a shuttle mission.

We knew that during Jerry's increment, the Russians failed an attempt to redock a Progress cargo vehicle to the Mir. All we heard was that this was an experiment to test some portion of a manual docking system. The Russians referred to it as a "TORU test." We knew TORU as the system the crew had used several times to complete the terminal phase of docking when the automated system had problems. Although we asked for more information, they refused to provide it. This did not raise a concern because most critical systems, including TORU, are tested routinely.

Jerry told us almost nothing about the test while he was on orbit, and because of his distant relationship with his crew, he could offer little more information during the debrief after he landed. On board, his crewmates had not volunteered the information, and Jerry had not pressed, leaving him to watch events unfold without understanding them. After the test, he did seem to be somewhat concerned about it, but his sketchy description left us with little basis for requesting more from the Russians.

Frank Culbertson and I only knew that the test was unsuccessful. Had we known the real plans for the test, we would have demanded far more information on plans, training, and safety measures. Since we were led to believe it had been a routine test to ensure the system was working as designed, we did not press for details. After hearing Jerry's concerns, Frank used the opportunity to ask Ryumin to be notified about future tests ahead of time, but Ryumin refused. The only sliver of new information he offered was that this test had been aborted early when the video system had failed. We had no reason to suspect anything out of the ordinary and no basis for insisting on more information.

We later learned that the test reflected the fact that the Russians had been dealing with a serious problem. We knew that many of the electronics for the automatic rendezvous and docking system were made in Ukraine, but we did not appreciate the depth of tension between the two countries or how deeply Ukraine resented the Soviet Union, especially Russia. With the dissolution of the Soviet Union, Ukraine had raised the prices on the hardware. The Russians resented the price increase and feared an eventual cutoff of the supply.

Russia responded to this problem with two initiatives. First, they began work to replace the critical Ukrainian systems with ones made in Russia. While this would end the threat of Ukrainian blackmail, budget limitations meant it would take years to complete. Until those new systems were available, the Russian space program was at risk.

Their second initiative sought to dock the Progress without the Ukrainian systems. A sharp flight controller proposed a scheme where the ground would command the Progress vehicle close enough to the Mir for the crew to take over final guidance with the manual TORU system. This required the ground controllers to get the Progress in the right position and at the right velocity so that the crew could safely complete the approach. The TORU system gave the crew control of the maneuvering thrusters and included a television camera for visual information similar to what they got while flying to the station on a Soyuz capsule. The crew would then have to maneuver the Progress using only visual cues, without the information that normally came from the Ukrainian radar.

This was extremely difficult. Anyone who has navigated a boat across a large body of water knows how hard it is to judge distance and velocity without the visual cues provided by landmarks in between. The same problem exists in space but to a much greater degree. That is why the TORU was only designed to be used after the automated system had positioned the Progress close to the Mir using the Ukrainian radar data for distance and speed.

This manual docking test would explore whether ground control could get the Progress where it needed to be and whether the crew could manage the rest of the approach. If successful, the crew would complete the rendezvous using only visual information—no radar. The crew needed to be well-trained because they had little margin to recover from any errors.

The Russian engineers made an honest effort to make this work. They provided the crew with a crude tool for estimating range by overlaying a grid on the television screen that displayed the Mir as seen from the Progress. Comparing the size of the Mir image with the grid would give an estimate of the range and could, theoretically, be enough. Ground testing in simulators had shown that if the Progress started the maneuver from the right place and at the correct velocity, if the camera provided a clear image of the Mir, if the crew remained properly oriented, and if the Progress and its engines performed as expected, a series of timed braking burns could guide the Progress to its normal parked location about fifty meters behind the station. From there, the crew would perform a routine manual docking.

To mitigate the risk of collision, the Russians designed the approach trajectory to pass by the station if the crew was unable to control it. If the braking burns were not performed, the Progress would

---

Manual Rendezvous

The proposed manual rendezvous and docking was a very difficult task for several reasons. First, long distances are extremely difficult to judge visually across the void of space due to the lack of reference points. Second, Earth's movement in the background made it difficult to judge the relative motion of the vehicles. Third, orbital mechanics often cause movements that defy normal intuition and require action long before the effects become apparent. Fourth, the low thrust of the Progress engines required long, sustained firings, so it took a long time to determine their effectiveness.

---

miss the Mir by at least a kilometer. This apparently happened in the attempt during Linenger's increment when the crew was not able to get the television picture to appear and, therefore, made no braking burns. The basic idea for the test was sound, as were many parts of the test procedure, but the final plan had serious deficiencies.

We never learned why the television picture was not available during that first test, but the Russian engineers blamed the lack of pictures on interference from the radar. This explanation was plausible, if not likely, but what happened next was inexcusable. Russian management decided to try again with the next Progress vehicle, during Mike Foale's stay, without understanding what had happened the first time and without reassessing the crew's readiness despite the many months since they had trained on the system.

The Russians did not share their plans with us, so we had no opportunity to consult or concur on the test. Jerry did not know what was going on at the time and only shared his vague concerns after his mission was over. By then, we could not get information on an event that had happened months before. Because we had an incorrect understanding of the test, we had no concept of the risk it involved.

Altogether, the first test was a colossal failure to communicate and think ahead. The Russians handled it poorly, and we could have been more aggressive in pursuing the issue when Jerry finally told us that even the cosmonauts were concerned. We probably would have pursued it if we had more confidence in Jerry's account. But he'd displayed no particular urgency in bringing the information to us, and we'd doubted that it had been as serious an issue as he later proclaimed it to be.

Our skepticism about Jerry's concern was reinforced by his debriefing on other subjects we knew more about. The one that got my attention focused on his extravehicular activity (EVA). Jerry was the first American to perform an EVA out of the Mir and the first to use the Russian spacesuit. Because of my interest in EVA, I attended his debrief.

The meeting was held in a conference room filled with crew and EVA experts. All were familiar faces from my work planning Freedom maintenance, and I expected to hear the crisp, businesslike back-and-forth I always heard following an American EVA. NASA debriefings were designed to give everyone a full understanding of what had taken place, and the engineers, trainers, and other crews actively participated in mining that information.

But Jerry's briefing was nothing like those meetings. Because nobody in the room was an expert in the Russian systems, Jerry was able to give a much looser description of the events, waxing melodramatic and exaggerating many points. He had a lot of fun hamming it up for his audience, and they enjoyed the entertainment value of what is normally a very serious business. That bothered me, especially because some of his observations were completely out of sync with what I understood about the Russian hardware and way of doing business.

He made some excellent observations that were important for future operations, such as how much harder it was to work face down on the outside of a spacecraft compared to working from a platform surrounded by familiar structures inside the cargo bay of the shuttle. This environment gave the shuttle crew stable workstands, helpful ways to move around, quick access to the airlock, good references for orientation, and even a sense of security. It also provided bright lights so they could keep working when the shuttle was on the night side of the Earth.

In contrast, Mir modules were cylinders with many protuberances sticking out at odd angles, leading to disorientation and even vertigo. Moving around was a tedious process of going hand over hand using handrails along sometimes circuitous routes, with the safety of the airlock relatively far away. There were few external work lights, so when the Mir went into the dark of night, the crew had to wait patiently until the sun reappeared forty-five minutes later. They also got very cold.

These were important lessons for ISS and the extensive EVA it would require. But Jerry's melodramatic delivery and emphasis on trivia cost him a lot of credibility with the experts. I was particularly annoyed as he repeatedly talked about "razor-sharp edges" on solar arrays that I had inspected and knew to be perfectly safe. When I asked one of our most experienced EVA experts if I was being too critical, he said, "Naw—that's Jerry being the ham, just playing the room."

This added to my doubts about what had really happened on his mission, although I remained confident that we had fully addressed the most critical problem he had faced—the fire and its aftermath. The other issues that we knew about appeared far less urgent, and none seemed to pose a significant safety risk to him or future crews.

Meanwhile, on board the Mir, Mike Foale settled into what we all hoped would be a very productive mission. We were optimistic, especially since Mike had gone to great lengths to become part of the crew. He had learned to speak Russian fluently and spent lots of personal time with his crewmates. Once on board, he joined in to help with tasks that had never been shared with non-Russians. It was looking as though we finally had a crew that was "in the groove."

That hope was shattered at 4:45 a.m. Houston time on June 25, 1997. A second Progress docking attempt had failed, and the Mir crew found themselves fighting for the survival of their station.

The test procedure once again required the crew to see Mir through the Progress television camera from a distance far beyond its use in normal operations. Since the Russian engineers argued that the first attempt had failed because the television signal had been blocked by electronic interference from the radar, the radar was turned off for this attempt. The crew got the television picture they were looking for, so the test continued.

But the decision to turn off the radar deprived everyone of information critical to the safety of the test. Had the radar been on, the crew would have been able to compare the range and closing velocity with the plan and determine whether the procedure was providing a safe rendezvous. Without radar, the crew did their best using only very limited visual cues.

Everything about that day conspired against them. The commander was exhausted due to poor sleep, triggered in part by a sleep experiment he was conducting for the US. Minor variations in the trajectory of the Progress caused it to approach from a slightly different angle and at a higher velocity than planned. This particular Progress also had unusually weak braking thrusters, so when the crew executed the preplanned braking maneuvers, they were less effective than expected. Without radar, the crew was unaware of the reduced performance. The commander—exhausted, minimally trained, lacking critical information from the radar he'd been trained to use, and under great pressure to complete the test—was unable to cope.

The result was that the Progress hit the Mir, severely damaging a solar array on the *Spektr* module before gently bumping along the station and drifting away. The force of the impact tore a small hole in the hull of the *Spektr* module, likely where the solar array gimbal was mounted, causing the atmosphere to leak from the cabin. Almost simultaneously, the crew felt the impact and heard the alarm signaling depressurization.

Depressurization is one of the most critical hazards a spacecraft crew can encounter, presenting an immediate risk of death or loss of the station. Both NASA and Russian crews were well-trained to handle this emergency. Each crew member had a role to play and reacted immediately.

After the collision, Mike Foale was sent to prepare the Soyuz for an immediate departure if the leak could not be controlled. The commander, Vasily Tsibliev, used a clock and manometer to measure the leak rate and determine how much time they had to locate and stop the leak. The flight engineer, Aleksandr Lazutkin, was assigned to find the source of the leak. He had glimpsed the Progress as it had passed by a window and had a strong suspicion that it had struck the *Spektr* module, so he headed there to inspect the damage and close the hatch if necessary. If the leak could be stopped, they would do so. If not, they would scramble into the Soyuz capsule and evacuate to Earth. Depending on the size of the leak, the available time could range from days to mere seconds.

Cluttered Hatches

Anyone who has seen pictures of the inside of Mir will have noticed hoses and cables passing through the hatch openings. This was the Russian way of connecting the systems across the modules to limit the number of connections that had to be made during EVA. While this decision simplified assembly, it greatly complicated the task of closing hatches in an emergency. Due in part to this accident, the use of such internal connections was severely limited in the ISS design.

As soon as he approached the *Spektr*'s hatch, Lazutkin heard the air escaping and began clearing the opening to close the hatch. Because the Soyuz was docked near the *Spektr*, Mike was close by and joined him as soon as the Soyuz was ready for departure.

The two worked feverishly to clear the hatch of its electrical and data cables and ventilation hoses. The major electrical cables had quick-release fittings located on a panel adjacent to the hatch and could be cleared within a few minutes. The air ducts were also easily disconnected, but one communications cable had been added later in the program and did not have connectors nearby. After searching unsuccessfully for the connector, Lazutkin attempted to cut it with a knife and got a small shower of sparks for his effort. Foale traced the cable to a connector that he was able to release, and they finally closed the hatch.

With air no longer leaking from the Mir, they did not have to make a hasty evacuation in the Soyuz. However, much of the electrical power for the Mir came from the *Spektr*'s solar arrays. With those cables disconnected, the station's batteries began to drain, and higher-level functions like attitude control started to fail. Without attitude control, Mir had power only when its arrays were in sunlight and happened to be pointed toward the sun. The slow tumble of the station prevented the batteries from fully charging, and during night passes, the station became utterly dead—filled with darkness and silence, punctuated by the creaking of the structure as it cooled.

No power meant no life support either, and the lack of air circulation caused humidity and carbon dioxide to build up in some areas. Foale later told me about floating in the dark, silent spaceship, worried that he might asphyxiate in his sleep if a bubble of carbon dioxide formed around his head. He soon discovered that while such a bubble did form, the buildup of carbon dioxide caused his heart to race and wake him up before he suffered any harm. He found it disconcerting nonetheless.

The crew's quick action had saved Mir for now, but there were months of work ahead of them to restore anything resembling normal operations aboard the outpost. On the ground, the repercussions of the accident were quick to appear.

Our lives would never be the same.

17. CONSEQUENCES

I'd expected June 25 to be another day of making Phase 1 more efficient, scientifically productive, and valuable to the ISS. But just before five o'clock in the morning Houston time, Frank Culbertson was awakened by a call from John Curry and James Medford, our systems experts in the Russian mission control center. They excitedly filled Frank in on what they knew about the Progress collision, giving him a running account of events even as the crew fought to close the hatch. As the communications pass ended, it looked as though the crew had stopped the leak and would not be evacuating. Beyond that, we knew little.

As soon as he hung up with them, Frank called me to begin what would be the longest day of our time in Phase 1. By the time I got to the office, Frank had made the initial notification to our management and was getting the latest update from our team in Russia. The news was a mix of good and bad: The leak had been stopped, but the station was losing its systems as the power failed; the crew was safe, but the future of the Mir was uncertain.

Frank and I immediately reverted to our familiar roles in managing yet another crisis. He got on the phone with headquarters while I began calling experts and managers to muster their help. With the crew staying aboard, at least for the present, we focused on learning what had happened and offering support. The Russian flight control team was overwhelmed, so we stayed out of their way.

Over the next few hours, we were flooded with support from colleagues eager to help. My office once again filled with experts as we began to get a picture of the situation on orbit. Many details were missing, but the team was better informed than they had been for the fire. At first blush, the mishap looked like a program-killing catastrophe.

Our critics did not need details before blasting us. Those who had been skeptical of our work from the beginning proclaimed our foolishness based on no information. Some were still angry that we had invited the Russians into "our" program, while others thought we should have ended our presence on Mir after the fire. Many called for an immediate withdrawal from Mir and reconsideration of the plan to include the Russians in ISS. While those were legitimate subjects for debate, they were distractions that did nothing to address the emergency in progress.

That was our job, and we gave it everything we had. Frank was on the phone constantly while I was digesting the developments in Russia and on orbit. We were single-mindedly focused on supporting our Russian colleagues. Amid utter chaos and scathing criticism, we stood side by side with them as they clawed their way back from the crisis. Our steadfast support both defined and amplified the value of the program. Although no sane person would have planned it this way, the crisis allowed us to prove ourselves as valued and trustworthy partners, and we did. It still amazes me how few of our American colleagues saw the value of our response.

Each crisis had grown our understanding of Russian systems and operations, and each offering of support had earned us an increment of respect from the Russians. After the fire, our system experts became more knowledgeable about how the hardware worked. Many of them paid attention as the events of the spring showed how the Russian systems worked and how our colleagues operated them. They were chagrined that they had previously ignored the obvious.

Now, our main task was simply to help the Russians preserve the station hour by hour. Although Ryumin hoped to someday return the station to normal operations, the immediate need was far more basic—life support and attitude control. Meeting these elementary needs kept them working around the clock. Our first contribution was providing communications passes through our ground sites, which were now working well, but we wished we could do more.

In the first hours after the accident, Ryumin and Blagov occasionally sent us updates on their work, but they were few and not very informative. This made it hard for Frank to satisfy headquarters and the clamoring press. By the end of the day, the critics seemed to be winning the media battle.

Once again, the greatest source of light for me was the courage and tenacity of the Russians. I had already learned that their culture valued a different form of courage than the popular American image. While everyone can celebrate the momentary heroism of someone running into a burning house to save a child, Russians place great value on the courage to resist adversity over the long haul. We knew of their tenacity during the siege of Stalingrad during World War II, but this was our own personal demonstration. If there was any chance of success, they would not give up.

Our team in Russia had a front-row seat for this show of determination, and most found it deeply inspiring. They also felt a comradeship as the Russians brought the same dedication to Mir that NASA had shown during the crisis of Apollo 13.

Their tenacity was contagious; when I talked to our team in Russia, it was clear that they were not writing the Mir off either. At about two o'clock in the afternoon Houston time—eleven o'clock at night in Moscow—I got a call from Mark Severance, part of our staff in Moscow, asking if he should take the last bus back to the hotel or hunker down in our office in mission control.

"There is not much we can do here," he said, "but I am ready to stay if you want me to."

"Mark, I really appreciate that," I said. "I know how uncomfortable it is there, but I think it is important to have eyes and ears there to cover what is happening. It also shows we will stick with them through the difficult times. If you can, I really think you should spend the night."

My heart swelled with pride when he quickly responded, "Jim, I am glad to do it. There is no place I would rather be!"

He and flight surgeon Terry Taddeo spent the night taking turns monitoring operations and trying to sleep on a miserable excuse for a sofa in their team room. Later, they told me that the discomfort had been easily forgotten, overtaken by the thrill of watching history in the making. Once again, I was so very proud of the people who had come to support Phase 1 despite the warnings of their cynical peers. They were true patriots who saw our future and worked tirelessly to make it happen.

Back in Houston, things were less dramatic, though no less intense. Late in the evening, as the day finally wound down, I went to Frank's office to compare notes. He updated me on the news from headquarters and his interviews with the press, and I shared the latest update from Mark. As I finished, I realized what a burden the day had been for him. Frank was a man of incredible energy and optimism, but on that night, his normally cherubic face was lined with fatigue, exhausted from dealing with critics and others intent on shutting us down. I tried to lift his spirits by showing that we had turned a corner.

"You know," I said, "I am starting to think there's a chance we can pull all this together."

My modest optimism came from working to solve the problems. I could already see the Russian flight controllers making progress and felt they had a good chance to save the Mir. More importantly, I saw our steadfast support through yet another crisis as the foundation we needed to sustain the ISS and the Russian-American partnership.

Frank's face reflected the other side of the day's events. While I had been wrestling with problems and seeing the rays of sunshine peeking out between the clouds, he had been caught up in the maelstrom of naysayers who wanted our efforts ended "before someone gets hurt!"

Weighed down with this tide of negativity, Frank could not see the sunlight.

"I am not so sure," he said glumly. "I think we might be done."

In truth, at that moment, it could easily have gone either way. But whatever the eventual outcome, we ended the day knowing we had done our very best to sustain our partners and the program. He and I had also supported each other seamlessly, and I thought we had done well.

The best news was that we were not concerned about the immediate safety of the crew. The Russian backup life support systems were performing well, with supplies that would last several weeks, and the Soyuz was always available to retreat to the ground if necessary.

The survival of the Mir was far less certain. The Russians would have to find a way to stop its tumbling so its solar arrays could generate power for attitude control. Only then could they restore the computers and other systems needed for habitation. The situation would be tolerable for a few days, but the clock was ticking, and solutions needed to be found soon.

That progress came slowly but steadily. Within a few days, the Russians had stopped the tumbling and begun to restore critical systems. Ryumin proudly shared their plans to find the source of the leak and restore the module to full function. Suspecting the leak was at the base of the solar array, they had people working on a plan to remove the solar array, cover the motor drive with a plate, and fill the surrounding volume with an expanding sealant. Hearing their plans, our engineering team shared their experience with a sealant that worked in the vacuum of space.

This cooperation among the ground crews helped reduce their stress, but there was nothing anyone could do to relieve the burden on the Mir crew. They bore enormous, relentless pressure not only from the work to be done but also from the guilt they felt about the accident. The pressure was felt most acutely by the commander, Vasily Tsibliev, who had been at the controls when the accident had happened. It grew further when some on the ground began criticizing his actions without knowing exactly what had happened.

My biggest concern was what the accident said about Russian decision-making. Up until this point, we had trusted that they would maintain an acceptable level of crew safety, but now we had ample reason to question whether their standard was sufficient.

At a minimum, we needed to gain greater insight into

> **Foale's Contributions**
> Mike Foale finally achieved our goal of becoming an integral part of the Mir crew. Not only did he perform perfectly during the emergency response, but he also made countless contributions throughout his stay. When it was time to stabilize the Mir, he assisted Tsibliev in using the Soyuz thrusters to stop its tumbling. This required converting the Mir's attitude information to that of the Soyuz. Mike was able to make those conversions in his head and help Vasily perform this critical task.

and influence over Russian operational decisions. Ryumin must have seen this coming and started to show some flexibility, though it was not easy for him. He was used to great autonomy, and sharing power was difficult. Many of his colleagues urged him to resist our requests, feeling that if we didn't like the way they did business, we should quit the program.

Ryumin knew he could not afford that luxury. His program was all but bankrupt, kept alive only with our funding. If consulting with us was the price of that funding, he would have to accept it.

My wife recalls the morning after the collision when I got up at four o'clock, unable to sleep. She urged me to come back to bed with the promise of a comforting snuggle.

"I can't sleep; I might as well go in and do whatever I can," I told her from the edge of our bed. "Yesterday was hell, and it's going to be hell for a long time to come. But I've been thinking all night that something good could come of this. For a long time, I've been trying to get both sides to put aside their selfish interests and work together. As terrible as this all is, it might just be the thing that finally forces us together. I don't see any other way to get past this."

That became my focus, and I watched carefully for opportunities to pursue it. They appeared quickly, beginning when Ryumin asked for our help flying repair hardware that did not fit on a Progress. We'd expected that, but he went on to ask for an American crew to support more EVAs.

We were surprised and delighted by this request. It proved that Ryumin was beginning to accept us as members of his team. Mike Foale was the perfect test subject; he had done one EVA from the shuttle and had trained in Russian EVA as Linenger's backup. Although he was not scheduled to do one on his mission, we knew he would jump at the chance.

But Ryumin wanted more. He wanted us to provide EVA backup for all future missions as well. He told us that the repairs could take many months and many spacewalks. He pointed out that the next American scheduled to fly, Wendy Lawrence, was not able to do an EVA in a Russian Orlan suit and asked us to consider flying her backup instead.

Wendy, a former navy helicopter pilot, was superbly qualified technically but too small to operate some of the emergency controls mounted on the front of the Russian suit. We agreed to consider moving up her backup, Dave Wolf, who was EVA capable, and sent our team off to assess what it would take.

They quickly came back with the answer that it could be done if Dave would focus on EVA training at the expense of training for his research. Dave was eager to take on the challenge; he could see history being made and wanted to be part of it. When Wendy graciously agreed to step down, we made the swap.

The next few months were a flurry of activity to help the Russians. On two internal EVAs, crewmembers entered the *Spektr* to recover our experiment hardware and restore most of the power that had been coming from *Spektr*. They did this by installing a plate covered with electrical bulkhead fittings in the hatch and using it to reconnect the power cables through the hatch. The plate had arrived on the last Progress and was installed immediately.

The crew did everything they were asked, often under enormous pressure, but they were exhausted. This increased the risk of errors, and sure enough, one evening, someone inadvertently disconnected a power cable that caused the Mir to shut down yet again. The crew was at their breaking point.

The next day, one of the cosmonauts developed a heart dysrhythmia, and Ryumin called to ask if we could accelerate the shuttle flight to bring him home. The gentler reentry in the shuttle would be less stressful than returning on the Soyuz. We were eager to help, but it could not be done in time to meet their needs. Meanwhile, the crew was ordered to rest, which resolved the heart problem. Cleared to ride the Soyuz, the Russian crew was soon replaced with two fresh cosmonauts.

When Ryumin asked for Foale to participate in an EVA to look for the leak, it produced a major collision between our safety cultures. The JSC safety team was extremely concerned about the lack of information about the hazards that had been created by the collision. Mir had no external cameras, so the crew would be exploring the damage blindly. NASA management made it clear that the decision of whether to let Foale participate fell to our office.

Actually, it fell to me. Incredibly, in the middle of this existential crisis, Frank left for a three-week seminar at Harvard. This was completely out of character for him, as was the fact that he did not respond to my emails or phone calls. I was on my own, solely responsible for NASA's approval of the EVA and all other program activities. George Abbey and others would kibitz, but the decisions were mine alone.

If I chose to allow it, Mike would accompany the new commander, Anatoly Solovyev, to search for the damage. Solovyev would approach the site while Mike provided support, but both would be near the damaged area.

The risk was real, but the only way to assess the damage was to let them inspect it. My NASA safety folks fought this because it left the crew to make their own decisions about what was safe. With the NASA safety representative to the Mission Management Team (MMT) adamantly opposed, I asked his boss, astronaut John Casper, for a formal recommendation. He agreed that there were times when we had to trust the crew's judgment.

I was glad for John's support, but I wanted to make sure we really understood what was going to happen and how the crew would respond to various problems that could be encountered. Before approving the EVA, I required that the NASA safety team brief the crew on specific things to watch for. Ryumin agreed, and a remote briefing was quickly arranged. Solovyev listened to the presentation, but during the EVA, he would do what he thought was right. Although some criticized my decision, I was confident for three reasons.

First, Solovyev was the most experienced EVA crewmember in the world and remains so at the time of this writing. By the end of that mission, he would have sixteen spacewalks on his resume, and I was willing to give him credit for knowing what he was doing. Second, there is only so much progress you can make by talking about hypotheticals; eventually, you have to go to the scene and take a look. Finally, this experience gave us invaluable insight into the challenges of joint operations on ISS.

I also knew that if I denied Mike's participation, the Russians would do it without him, and we would miss a valuable opportunity to advance our cooperation. I felt it was worth the small additional risk to have Mike outside, especially since he would not be in the most hazardous areas. He was also unwavering in his desire to participate.

The EVA proceeded as planned, though Solovyev was unable to locate the leak. While Russian engineers worked on a possible repair, the operations team focused on restoring normal functions.

Meanwhile, a new storm was brewing in Washington.

The Globe Swift that nearly killed me, 1973. Credit Jim Van Laak

Serving as an F-106 pilot, 1982. Credit Jim Van Laak

NASA Johnson Space Center. Credit NASA

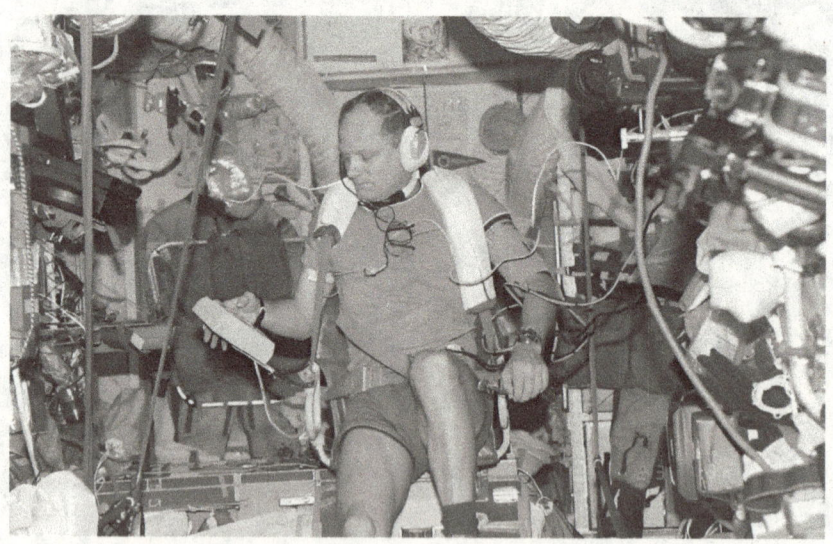

Norm Thagard on Mir. Credit NASA

STS-71 undocking from Mir. Credit NASA

My first visit to Moscow. Credit Jim Van Laak

Shannon Lucid on Mir. Credit NASA

Me and Frank Culbertson in Star City. Credit Jim Van Laak

Phase 1 Russians. Valery Ryumin and Yelena Kondakova, third and fourth from the left. Credit Jim Van Laak

Damage to *Spektr* module on Mir. Credit NASA

Seven special Americans—Phase 1 Crew. Credit NASA

ISS *Zarya* module and *Unity* node. Credit NASA

Star City with my interpreter. Credit Jim Van Laak

Visiting Buran shuttle with Dave Lengyel. Credit Jim Van Laak

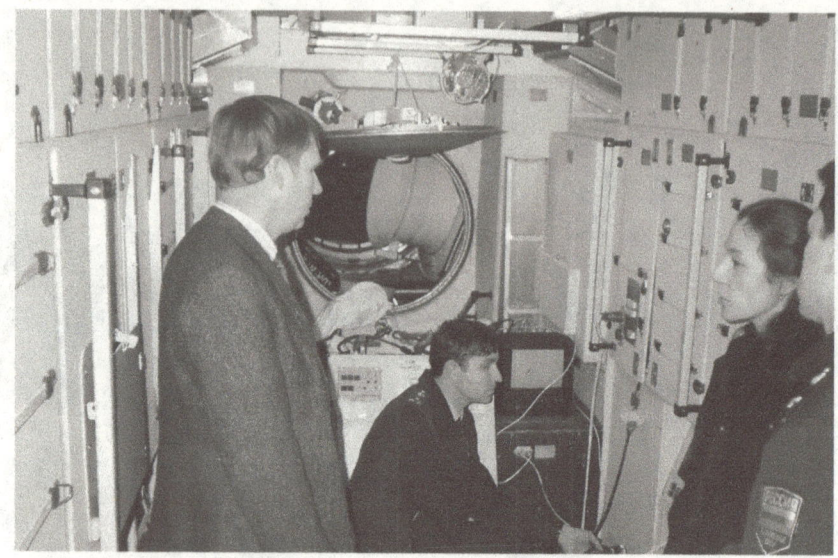

Visiting *Zvezda* trainer. Credit Jim Van Laak

Expedition 1 crew. Credit NASA

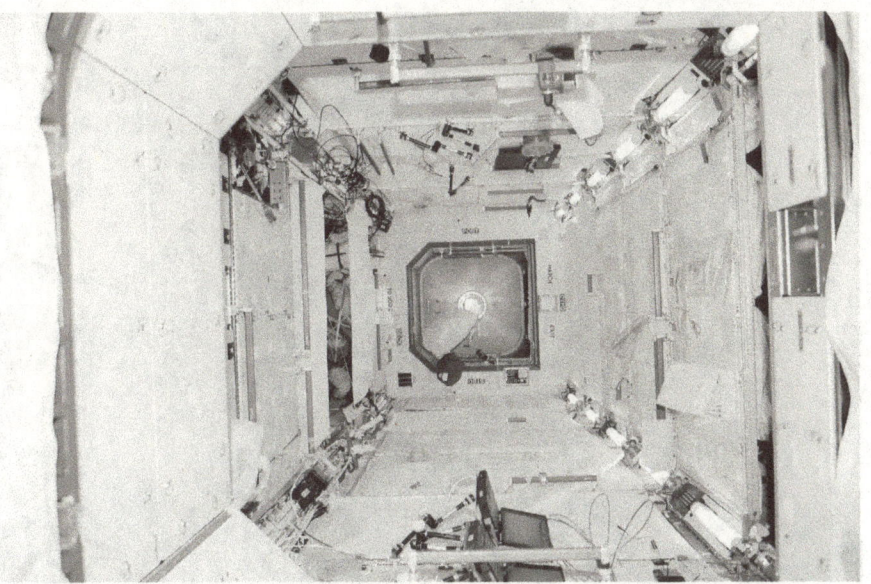

New *Destiny* module—that new car smell. Credit NASA

Destiny one year later—the lived-in look. Credit NASA

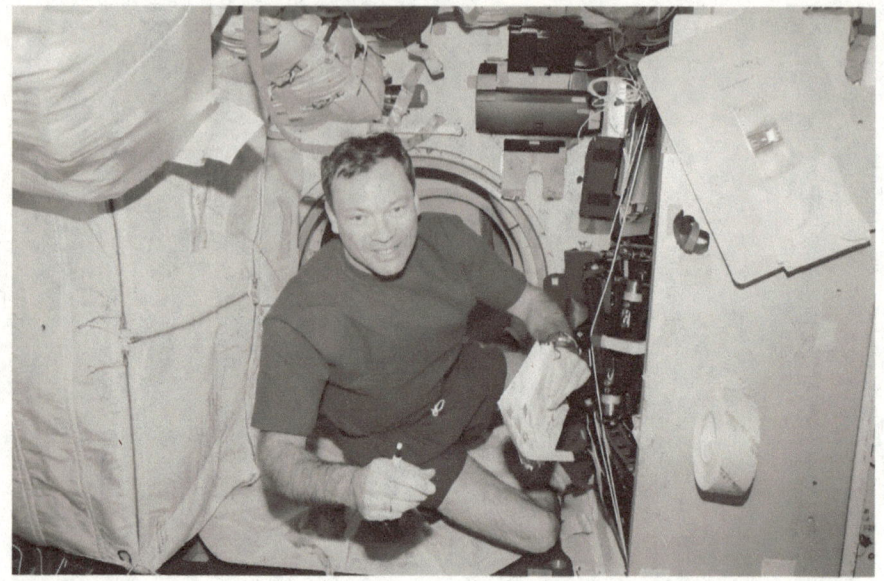

Cargo stowage on shuttle. Credit NASA

Chairing the MMT. Credit NASA

What NASA promised, 1993. Credit NASA

What NASA delivered, 2011. Credit NASA

The ISS receives the Collier Trophy. Credit Jim Van Laak

Mission Control—my home away from home. Credit Jim Van Laak

In shuttle trainer with my daughter, Cheryl. Credit Jim Van Laak

18. FALLOUT

The day after the EVA, as I watched the first rays of dawn spread across JSC, I wondered what I had gotten myself into. Our simple joint program to learn from our new colleagues had become an existential crisis for both of our space programs. How had this happened? What did it say about our efforts, their wisdom, and their value? Were we being courageous pioneers or stubborn fools?

These questions were becoming prime topics of conversation—not only within NASA but around the world. The collision was the third major crisis to hit the Phase 1 program within six months, and it captured the attention of the media. For weeks, the effort to save the station was a day-to-day struggle with an uncertain outcome. It offered the drama to captivate space fans and even interest those who were not enthusiasts. JSC was soon hosting reporters by the droves, and Frank Culbertson and I were in constant demand for interviews. We worked hard to keep rumors and speculation under control.

Our new visibility also brought oversight and criticism from all quarters, making us thankful for those who continued to support us. George Abbey, the JSC director, stood quietly behind us, offering whatever he could to help. Dan Goldin, NASA's administrator, had called us the morning of the collision with a crystal-clear message.

"I want you to know that when it comes to this issue, I work for you. Whatever you need, I will do my best to support. *I work for you!*"

While Dan's support was deeply appreciated, it was not the norm. Many people who had once supported the program now refused to listen to the facts. Some tried to talk us into shutting it down, while others pulled back to avoid being criticized if things got worse. The most distressing were those who encouraged us in private but criticized us in public.

An unanticipated problem arose in our relationship with the Space Shuttle program despite being led by our former boss, Tommy Holloway. The Space Shuttle program team was used to having total control of their work and had developed strict processes to guide it. They deeply resented being asked to compromise those processes to assist our much smaller program, especially when the problems were coming from another country. They continually pushed back on our rapidly growing list of requests to support the Mir. Instead of doing their best to help us, they grudgingly accommodated only what they considered to be our most serious requests while denying anything that did not meet their sense of priority.

I was astonished that they didn't see their fate as tied to ours. During the summer of 1993, the White House told us that there would be no station without the Russians, and without the station, the shuttle's future was dim. Their blunt warning had provided the urgent pressure to craft the partnership needed to enable ISS and, by extension, the Space Shuttle program. If we were successful, we would forge the team we needed, prove the value of ISS as a tool for international cooperation, and elevate the shuttle from a standalone launch vehicle to an element in a grand, global space program.

Likewise, our failure would be devastating. After all, if we could not make this program work when we were

> **Messages from Mir**
>
> In 1997, Peter Mass wrote an excellent piece in *The New Yorker* magazine about the Mir program and the political maneuvering going on in the background. During an interview, I related how "there are plenty of people within the political system and within NASA who are privately telling us to go, go, go, while at the same time distancing themselves from any blame."

operating mature vehicles supported by experienced teams, how could we hope to be successful in assembling something entirely new on orbit? Despite this compelling logic, the Space Shuttle program team felt no particular need to support us.

The tensions among the NASA team erupted one hot August day. The Russian plan for sealing the leak required a large plate that did not fit in a Progress. With Frank at Harvard, Ryumin called me to ask if I could carry it on the upcoming shuttle flight. I agreed to see what could be done.

The Space Shuttle program offices were just down the hall in Building 1, so it took mere moments to walk down to the office of the program's deputy manager, astronaut Dick Richards. I explained what Ryumin had requested and why it was important, making it clear that I supported the request. Although this would mean extra work for his team, its importance seemed obvious.

"It's way too late for such a big change," Dick said crossly, "especially when it's a result of their own damn screwup."

He flatly refused to accept the plate in the shuttle cabin. Holloway was not available to hear an appeal, so I went back to my office and called Spacehab to see if they could help. I would only elevate the issue to higher NASA management if they could not.

"Sure, we'll take a look at it. If it fits through the hatches, we should be able to make it work," said Mike Kearney, Spacehab's president. The difference in attitude between the complacent bureaucracy of the Space Shuttle program and the hungry contractor was breathtaking.

At nine o'clock in the morning, I went to our weekly tag-up with the JSC director, George Abbey. Astronaut John Young sat at the other end of the table, quietly taking notes. After reviewing Foale's status and the work to repair Mir, I mentioned that Spacehab would be taking the plate because the shuttle team didn't want to mess with it.

"Oh, why's that?" George asked in his usual voice, so quiet it was barely audible across his narrow table.

"They hate late changes like this because they wreak havoc on their neat stowage plan and require new analyses. Their process assumes NASA has control over everything and can confirm our needs months in advance. Dick is right about the new work, but he is missing the whole point. This is what it takes to support a space station. When bad things happen, we have to react as best we can. Who knows what ISS will require down the road?"

George nodded and silently wrote in his notebook, exchanging a sideways look with John Young.

An hour later, John stopped by my office.

"Tell me again why the shuttle won't carry your hardware?" he asked.

"Our request is coming in very late to a process designed to freeze everything months in advance. Shuttle wants everything perfect, and that works great when you can afford the time, but things have changed. Now we are supporting an ongoing mission, and they hate having to react to events instead of controlling them. They also don't like doing things for the Russians."

John said nothing, listening intently. I went on.

"But in my view, none of that matters. This is our future and theirs. The decision to build a space station was also a decision to use the shuttle to support it in whatever way is required. They are in a support role now and need to react to events and carry whatever emergency supplies are needed as quickly as safety permits. They need to accept that reality instead of fighting it."

John nodded and left.

About twenty minutes later, Dick Richards stormed into my office.

"Okay, okay, I'll fly your damned hardware! Just call off George's dogs!" Pausing only long enough to be sure I had heard him, he was gone. I allowed myself a silent smile.

As difficult as it was to convince our colleagues to change their ways, at least we could confront them with facts. Over the course of the summer, a wave of concern and objection had been building among those who had never liked our program. Everyone from retired astronauts to members of Congress had spoken out about why it was not worth the risk to stay involved in Mir. All of them had expressed concern for the safety of our crews, but none had taken the time to explore the risks or how we controlled them. They had ignored the fact that the most important outcome of our work was reducing risks for ISS, a far more difficult and ambitious program that would last decades and cost far more. I marveled that such strong convictions could be grounded in such ignorance.

One weekend in early September, after the Mir had suffered yet another minor setback, ABC News called JSC Public Affairs with an ultimatum. They declared that they were going to do a major story on Mir "whether you participate or not!" Since Frank was once again away, our lead public affairs officer, Rob Navias, came to me first thing Monday morning. His tone was somber.

"The mood in Washington is grim, so this could be a killer. We need you to show them that we have a plan to work through all this. It is important that you make yourself available."

Seeing Rob's obvious concern, I took pains to clear my afternoon schedule. Then, I called Moscow to make sure there were no late developments.

At about two o'clock in the afternoon, I walked to the media center through the Houston heat and humidity. Although still quite shy, I had been forced to overcome that by the flurry of media interviews over the summer, and I was comfortable with my task because I had a message. This was an important program doing something vital for America, and the recent problems, while serious, were being overcome. I was ready.

ABC's space reporter, Ned Potter, was in his office in New York while a news crew from the local ABC affiliate in Houston was with me. After a few minutes of coordination, the camera crew began to shoot as I answered questions coming from Ned over the phone. He was respectful and asked good questions, and I ended the interview comfortable that I had represented us well.

As I left the recording session, I was accosted by a camera crew and a reporter from another local station. They pleaded with me to give them an interview as well, and Rob's pained expression said I should.

The pretty reporter was warm and friendly as she set up with her crew, but when the camera came on, she became an attack dog. Such behavior is normal, but in this case, she had absolutely no idea what she was talking about. This made the interview uncomfortable because there was no way to overcome her ignorance without making her look like an idiot. Although I might have enjoyed it, it would not have been helpful. Instead, I did my best to answer her questions in an honest and useful manner while she ignored my careful answers and kept hammering away at the idea that we were acting foolishly without considering the risks. Finally, in frustration, I gave it to her straight.

"Look, it all boils down to this: Flying people in space is hard and dangerous. We do our best to plan ahead and manage the risk, but we can't anticipate everything. If people can't deal with that fact, they should crawl back under their beds and forget about having a space program!"

Rob rolled his eyes but said nothing. Naturally, that clip was what made the local news, and it was a hit with my colleagues. They loved the image of timid souls cowering under their beds while the adults did what needed to be done.

The ABC story was less successful.

As was typical, I worked in my office until late. Anticipating the story, I watched ABC News on TV while I reviewed safety reports for the next shuttle flight. As the show wound down with nary a word about Mir, I began to congratulate myself for doing such a great job that there was no story to report.

Then, in the closing seconds of the show, anchor Peter Jennings picked up a piece of paper and announced, "This just in. We hear the Mir is spinning like a Frisbee." With that, he closed the broadcast.

I was stunned. Never before had I encountered such a complete fabrication by a major network. It took a moment to recover from the shock of what I had just seen, after which I picked up my phone and called directory assistance in New York City to ask for ABC. When I got their operator, I asked for Ned Potter. Seconds later, a now familiar voice answered, "Ned Potter here."

"Ned, this is Jim Van Laak in Houston. What the hell happened to the story?"

"Jim, hello. I don't know. I didn't listen to the broadcast."

"Well, it was a disaster. Why don't you toddle down the hall and find out what happened? Then please call me back."

Taking my number, Ned agreed to investigate. About twenty minutes later, he called.

"The tape was in the machine and ready to go, but they were a little behind schedule and decided to skip it."

"What about Peter's statement?" I asked.

"He pulled that out of his ass." Nothing more needed to be said.

Others were on our case as well. Congressman James Sensenbrenner of Wisconsin had never liked the plan to bring the Russians into the station program and had been the driver behind our need for the contingency plan for the Russian elements—the plan I had delivered to Goldin four years earlier.

With this episode, he turned his righteous fury on us. He called for an investigation into the collision and the whole issue of safety on the Mir. We were soon inundated with demands for information and records. Between his staff, the General Accounting Office (GAO), and the NASA inspector general, we were required to send forty-three large boxes filled with copies of our emails, memos, technical documents, and anything else relevant. Even our personal records, like the notebook I took to meetings, were copied and sent along.

More distressing was the tenor of the conversations. Some organizations and individuals, like the GAO, were professional, while others seemed to want to intimidate us. Even our friends added to the workload. Although Dan Goldin trusted us, he needed to exercise due diligence and requested an independent review to ensure we were making good decisions. Despite the extra work, we genuinely appreciated the chance to go over our logic and decisions with someone both independent and qualified to understand the situation. He tapped Tom Young, a highly respected aerospace manager, to take a fresh look. After a thorough review, he concluded that we were doing a good job balancing the risks we faced.

Others were less professional. One astronaut, Blaine Hammond, went around telling people that we were being reckless with crew safety, which he concluded without having exchanged one word with us. We later learned that he had met privately with the inspector general and accused us of gross negligence. He'd also shared his views with Tom Young, who'd listened carefully, weighed every point, then sided with us. Still, Blaine remained ignorant and critical.

Through all of this, we continued our preparations for the next shuttle flight, STS-86. This mission was critical to keeping the program on track. It would exchange Mike Foale for Dave Wolf and carry up a fresh set of logistics and critical repair materials for the Russians.

As luck would have it, the Mir's age started showing itself in a new way about two weeks before launch. The computers that provided attitude control were beginning to fail intermittently. But this time, everyone on both sides responded quickly and openly, allowing us to craft a unified response. The Russians told us everything about the failing hardware and its impact, then scrambled to get a spare computer to KSC. Spacehab jumped in and found a place to stow it a few hours before launch. The Space Shuttle program developed a special procedure to allow the shuttle to dock if the computer failed at a critical moment. We coordinated the work and kept the pieces working together. It was a strong team effort, and I was thrilled to see it come together.

During the flight readiness review at KSC two weeks before launch, a courier delivered a letter to Frank. He read it, then gave it to me. It was a formal request to appear at a hearing in front of Sensenbrenner's subcommittee one week later, the midpoint between the review and the launch. Headquarters told us it had the legal authority of a subpoena, so after the meeting, we flew back to Houston and added preparation for the hearing to the already-heavy workload of the upcoming mission.

The staff of our tiny office, helped by volunteers, stepped up to the challenge. One of the most helpful was Missy Gard, a NASA employee from Marshall Space Flight Center who was on a detail at NASA Headquarters. She offered to come to Houston and help us organize the chaos. She would eventually move to Houston and become one of my most valued colleagues in ISS.

The team worked long hours to pull together the briefing books Frank and I would have to master before the hearing. Although I felt somewhat intimidated by the prospect of testifying in front of Congress, by working straight through, we felt we had our story together. We flew up the day before the hearing, with Frank going early so that he could speak to members individually ahead of the event. Most were pleasant and professional, but the chairman was hostile, using profanity and bluster in an effort to intimidate.

Early on the hearing day, Frank and I met at NASA Headquarters for a final strategy briefing before the big event. Frank was remarkably calm as we were driven to the Capitol and led to the hearing room. As we walked in, I mumbled under my breath, "Democracy in action."

One of the cameramen standing nearby overheard my comment and corrected me. "Not democracy," he said. "Theater." How right he was!

Frank and his fellow panelists were introduced, and introductory remarks were made. I sat directly behind Frank, armed with reference books and slips of paper to feed him any details he needed. Frank had an amazing mastery of details and did fine on most everything, with my contribution mostly limited to occasional suggestions of something he might want to add.

Midway through the hearing, during a break to allow the members to vote on a bill, Ned Potter came over and introduced himself. He apologized for what had happened a few weeks earlier and promised me better treatment "next time." I quietly hoped it would not come.

The hearing resumed. Each member had five minutes to ask questions and get our answers. In practice, hostile members mostly gave long speeches for the cameras, usually irrelevant to the issues. Those who asked questions generally did so at the end of their speeches, leaving us with only seconds to answer.

The hearing, which had begun with a roar of indignation, ended with a whimper. Fewer than half of the members were present, and only a handful attended the entire hearing. Even Sensenbrenner had left after his opening comments berating NASA and returned only to wrap up his attacks. He clearly had no interest in learning the facts.

Having survived the hearing, we raced back to our real jobs. After days of hard negotiations with our Russian colleagues, on the night of the launch, I escorted them to the launch site as usual. As we rode to the viewing area, I got a call from Jeff Bingham, an experienced congressional staffer then serving on a detail with our congressional liaison office.

"Sensenbrenner just told the media that he would hold you and Frank criminally liable if anything happens to Wolf. Should I call Frank and tell him?"

"Frank is busy in the Launch Control Center. The last thing we need is for the operational team to be interrupted with congressional grandstanding. I'll tell him later."

When I did so after the launch, he shook his head in amazement. "What is wrong with these people? They want a grand and glorious space program, but they're unwilling to take any risk at all. How do they think we got this far?"

The shuttle mission, STS-86, went well as usual, but our smooth professionalism did nothing to stop the criticism. Most of the complaints were about Dave's reduced research program, but research had always been a secondary reason for our missions to Mir; our goal was to prepare our nations to work together on ISS, and that made our choice of Dave's priorities clear.

A few weeks after the shuttle's return, I got a call from Jim Oberg, a former NASA flight controller who was working as a journalist covering the Russian space program. The Soviet space program had been a lifelong focus of his attention, and before the fall of the USSR, his expertise had probably been unique in America outside of the CIA. He had been one of very few Americans to visit the launch site at Baikonur and talk to the technicians there. Now, he was an independent journalist writing about the space program.

Jim had called to share a rumor about the condition of the Soyuz that had returned Tsibliev's crew in August. His source told him that so much condensation had accumulated in the Soyuz during the power outages that it had shorted a connector and caused the braking rockets, which softened the touchdown, to fire when the parachute opened instead of just before touchdown. The source also said that if the rockets had fired when the heat shield had still been in place, they would have burned through the hull and incinerated the crew. Jim wanted to know what I thought of this possibility.

"Well, Jim, that is the first I have heard of this issue, and to be honest, it seems extremely unlikely. But if true, it would be a big deal, so I'll find out."

I immediately called Victor Blagov, the senior flight director for Mir. He agreed that it seemed unlikely but promised to find out.

The next day, Victor called to tell me that he had researched the question of the rockets firing early. He admitted that they had fired early on this Soyuz, but the designers showed him that the rockets were absolutely fail-safe. Not only was it impossible for them to burn through the hull, but they had also done tests to prove that if the rockets somehow fired while on orbit, the whole landing system, including the heat shield, would work fine. This was the kind of robust approach I was used to seeing in Russian hardware.

But one of the most distressing surprises about this post-collision turmoil came nine months later as we neared the end of the program. While working on preparations for the final mission, I got a call from Bill Broad, a reporter for *The New York Times*. Although reporters usually went through the public affairs office, they sometimes called directly to get our unvarnished response to a pending story.

"How do you intend to respond to Blaine Hammond's letter?" he asked.

"What letter?"

"The one he gave to the IG."

"I have never seen it," I said with forced calm. By that point, we had been through so much that one more poison pen letter was not going to make much difference. "Could you fax it to me?"

Five minutes later, I held the letter Hammond had provided to the IG during the investigation the previous year. Although it contained the same stream of unsubstantiated claims that he had given to Tom Young and others at that time, it had never been provided to us so we could answer its points. I took the letter to Frank, then George Abbey.

"What bothers me," Frank said, "is that he should have come to discuss these concerns with us directly. If he had real information, we would have acted on it to make the program safer. Instead, he wrote this CYA letter that contributes nothing to NASA or the safety of the crew."

Indeed.

19. GROWING TOGETHER

Dave Wolf was not a part of the original cadre of Mir astronauts and only got his chance when the ISS launch was delayed by about a year. The delay had been announced long before the crises with Mir, when the magnitude of NASA's ISS work schedule had finally been understood. Around the same time, Russia had made it clear that they were taking longer than planned to finish the service module, later known as *Zvezda*.

Looking for a way to fill the gap, it was proposed to add two additional long-duration missions. Frank and I liked the idea of adding missions to build our team relationship, and the Russians were thrilled with the additional funding. When we began looking for two more astronauts for the program, I recommended Dave as someone eager to get back into management's good graces. He jumped at the chance and was ready to go when the *Spektr* crisis—the collision and the subsequent leaks—accelerated his flight. He was a great fit.

Dave was popular with the press, intelligent, and hardworking. But he was also quietly flamboyant, with a talent for getting in trouble. It was that awkward talent that led to my meeting him and ultimately to his flight on Mir.

Dave's combination of likability, outgoing nature, and many interests made him a natural target for people who wanted to get close to the space program. This led to him being drawn into an FBI sting operation in the early nineties. While Dave did nothing wrong, the

196

incident received a great deal of attention, and the media storm surrounding the event hurt his career.

Several years later, another of his flamboyant acts brought us together. He and another astronaut were partners in an aerobatic biplane, and one day I saw the plane perform an aerobatic maneuver over our community airstrip much lower than is allowed by the FAA. Because this was illegal and irritated the neighbors, not all of whom liked having an airstrip nearby, I made a mental note to track the plane's owners down and speak to them.

An hour later, the problem escalated. One of the officers of the homeowners association came to my house and demanded I get the pilot's name so he could be turned in to the FAA. While his response was justified, I was concerned that it could have deeper repercussions. NASA works hard to develop an astronaut corps with a carefully crafted balance of boldness and responsibility, and I worried that a public punishment could damage that balance. Explaining this to my neighbor, I asked that I be allowed to take care of it unofficially while guaranteeing that it would not happen again. I promised that the astronaut would get a truly world-class butt-chewing.

Winning my neighbor's reluctant agreement, I immediately contacted the head of the astronaut office, "Hoot" Gibson. Hoot was one of the most colorful and storied members of the office and one who himself had been the center of some controversy a few years earlier. I explained what had happened and asked for his help, emphasizing that the next astronaut to violate our neighborhood's airspace would receive no mercy.

Hoot got the message. He quickly identified Dave as the pilot and delivered the promised corrective action. Dave was suitably chastened and cheerfully cleaned up his act. Hoot even arranged for me to see Dave express regret for his actions and promise to stay out of trouble. I was pleased to see him admit his error, but the incident also demonstrated that he was unflappable—an important credential for working with the Russians.

That is why, after promoting Dave's move to Phase 1, I felt particularly responsible for him while he was on Mir. Dave was already in Moscow when the change was made, and he enthusiastically jumped into his EVA training. Rick Nygren's science team struggled to build a research program to fit his changed mission plan.

One of the most aggravating aspects of working with the Russians was their refusal to consult with us about resource allocations on Mir, which often caused serious damage to our research. Now that the Russians desperately needed our help after the collision, we made it clear that we expected to be treated as respected partners, not just their trucking company. The public outcry over the accident had brought an unprecedented level of scrutiny that even Ryumin could see.

Some Russians complained about our demands for a more balanced role in the program. They felt that we were demanding more resources as the price for our support at a time when Mir was barely surviving. But Ryumin now understood that we did not necessarily want more resources but rather to be a part of the decision-making process when the resources were divvied up. Waking up to this at last, he began to include us in their planning activities.

Ryumin also saw that public concern over the safety of our astronauts threatened public support for our work. Frank and I told him we could overcome this if we could show how our teams were growing together. We also shared our opinion that another such incident would certainly end Phase 1 and possibly ISS.

For us, this moment of real progress perfectly illuminated the value of Phase 1 in preparing for ISS. The lessons we were learning and the trust we were building were critical to the success of the new program. However, when we looked for evidence that the ISS management team was learning from our experience, we were deeply disappointed to find that most were only dimly aware of the accident.

When I challenged my colleagues about this, their answer was that Mir was irrelevant because "NASA will be in charge" of ISS, not the Russians. They assumed the Russians would follow our direction and ignored how dependent ISS would be on the Russian elements and operations. In Phase 1, we were working hard to develop a balanced, mutually respectful way of working together, but few on either side were ready to change their deeply rooted, parochial ways. It was embarrassing to see that, in some ways, Ryumin was ahead of some American colleagues.

We were also surprised to find that some Russians did not see the problems on Mir as being that serious. Yes, they'd had a few tense days, but to them, those were just one more bump in the road to space. They had suffered many challenges over the years, so they responded as they always had—by buckling down and working through the problems one at a time. They were shocked when some Americans called for us to quit the program after the collision. They saw these timid souls as cowards unwilling to face adversity. More than one Russian was heard wondering how they could work with people so quick to turn and run when things got difficult.

Once again, I saw wisdom on both sides. I deeply respected the Russian courage and dedication to their goals in space, but I also shared the American goal of learning from mistakes and not repeating them. One of my goals became finding a mutually respectful blend of Russian audacity and American caution. It would prove to be highly elusive.

Many on the US side were eager to help the Russians work through the problems the collision had caused. One of our contractors even came up with a novel idea to look for the source of the leak by injecting a gas into the *Spektr* that would fluoresce in sunlight as it leaked out. It could only be seen in photographs using a filter on one of our cameras and would have to be analyzed on the ground later. We decided to give it a try.

The effort appeared to go well. Shortly after undocking, the shuttle held position while the Mir's crew released the gas into the *Spektr*. As the gas vented into space, the crew took pictures while the shuttle flew a carefully planned path around the Mir. It was a beautiful bit of choreography with great cooperation by the crews and control centers.

Only later did Frank and I learn that the whole exercise had been in vain. At the last minute, and without our knowledge, a mid-level shuttle manager refused to carry the critical filter on the grounds that it was glass and could fracture into dangerous pieces. He ignored the fact that other lenses and filters were routinely carried with appropriate steps to control the hazard.

This horrific failure by the NASA team was made worse by the fact that all the people involved worked near each other in JSC's Building 1. If this manager had taken just a moment to consult with us, the problem could have been solved. Instead, his actions wasted precious resources and planted another seed of doubt in the Russians' minds.

The rest of the mission went well. Mike Foale came home to a hero's welcome, and Dave began his adventure on Mir. The outstanding cooperation in fixing the Russian computer failures had previewed greater cooperation between our teams. The Russians saw that we were genuinely trying to help them, which built our credibility.

While Dave's mission held no program crisis, it was still challenging for him. The change in sequence meant he also flew with a Russian crew he did not know well. John Blaha had faced this problem due to a medical issue with his crew, and Dave experienced it too because he'd replaced Wendy. He had little time to get to know Anatoly Solovyev and Pavel Vinogrodov because they'd launch on Soyuz more than a month before he launched on the shuttle. We worried about this, but there was no alternative.

The subject came up when I got together with Dave right before he headed to KSC for his launch. He was eager to restore his place in the JSC community and promised to do it right. If he had any concerns about the mission, he concealed them well, and I felt confident that we had made a good choice.

When Dave got to the Mir, he immediately jumped into the work, making himself a valuable member of the crew. Mike Foale had coached him on ways to ease some of his colleagues' burdens and make life a little easier for everyone. As a result, he volunteered to take over one of the most onerous tasks—cleaning up the condensation that continued to dog the vehicle. The work was unpleasant, hard, and dirty; Dave's decision to make it his responsibility was a huge relief to his colleagues.

The rest of Dave's increment was unremarkable. He did a good job for us, and I was glad to see him return home with a successful mission under his belt.

20. WRAPPING UP

Dave's time on Mir gave us the first stability since the collision six months before. Despite the constant string of problems, we were finally making progress.

The final increment of the Phase 1 program would be flown by Andy Thomas, a quiet and intense astronaut with joint US-Australian citizenship. Andy had joined the program as Dave's backup. Because Dave had originally been scheduled for the last mission, Andy had joined with little expectation that he would fly to Mir. He was surprised to move up to prime for the last mission but glad to have time to prepare. Although still compressed, his training benefited greatly from what had been done for Dave. He was a quick study and was as prepared as we knew how to make him.

As with Dave, Andy would spend the first months of his mission flying with a crew he did not know well. More importantly, he was bothered by their attitudes. Anatoly Solovyev did not care for the American space program or astronauts, and his negative attitude made everything more difficult. Andy was concerned.

Since I had not known Andy before his selection, I heard about his concerns through his ops lead, Scott Gahring, a fine engineer with whom I had worked on Freedom. Scott told Andy that I was a straight shooter and convinced him to share his concerns with me. Shortly before he launched, we talked for an hour about the history of the program, our experience with Anatoly, and especially what to expect in the early weeks of his mission. I told him about the entire history of our time with the Russian station and team, focused on why I thought it was safe, and urged Andy to share anything that might contradict that judgment. He had already talked with the other Mir astronauts, but each of them had gone through a unique experience. Our conversation was the first time those comments were put in context for him.

We also discussed Solovyev. Anatoly had spent a tour as a Russian liaison in Houston and often joined our telecons with Ryumin. Like most, I deeply respected his competence but was put off by his demeanor. His expertise with the Mir was unmatched, and Ryumin put great stock in his advice. He was also the perfect commander to execute the most difficult repairs, able to react quickly and effectively to any problems that might arise. But his arrogant and aloof attitude was a problem. He acted as if he were superior to us in every way. His behavior on the EVA with Mike Foale showed that he was willing to do whatever he felt was right, regardless of direction. He had gotten away with it, but I had no idea how much of a risk, if any, he had taken.

Andy was concerned but knew that such arrogance was common among crew members from both countries. He promised to comply with any sensible directions he got from the ground.

"I am a part of the team," he said, "and I will do my part."

"We appreciate that, Andy, but you will find that you will be on your own much more than when you flew on the shuttle. Because of the comm problems, we will not be able to support you as much as any of us would like. You will have to address some things without us. We know that and will support you every way we can, including backing your decisions when you have to make them. This is new territory for us, and you are a pioneer. Pioneers sometimes make mistakes, and we understand that. We will have your back!"

As we concluded, I promised Andy my personal support and respect for the challenge he was about to undertake. I was very glad for our personal connection, as small as it was, and glad that he was going.

The following week, we were at KSC for the launch of STS-89 and our usual meetings with Ryumin and his management team. At the last minute, we were all invited to the grand opening of the new facility that housed one of the Saturn V rockets left over from the Apollo program.

The event was a black-tie affair with VIPs from all around the country. Most of the living Apollo astronauts would be there, together with senior leadership from NASA and related industries. Once again, Phase 1 had been forgotten during the planning, but at the last minute, we were added along with our Russian colleagues.

The facility was gorgeous. One of my colleagues, Jim Snowden, was eager to show me his contributions to the markings on the hatch window that the crew could use to align their spaceship for entry in case their instruments were not working. As we studied the capsule, a familiar voice called out from behind.

"There he is, pointing out that doggone line again! Don't you have anything better to talk about?"

We turned to see Gene Cernan grinning broadly. Although I recognized him instantly, I had not met the last man on the moon. Knowing he was both Jim's boss and friend, I was thrilled to see their mutual respect. It was exactly what I hoped would follow my time in NASA. A few years later, I would enjoy the same rapport with this fine gentleman.

When I returned to the foot of the Saturn V, I found my Russian colleagues gathered in a tight group standing under the enormous engine nozzles. Valery Ryumin, his voice broken with emotion, was speaking reverently about this beautiful machine. Although his emotions were liberated by alcohol, it was nice to finally see this respect coming from such a proud Russian.

The next day, STS-89 got off perfectly. Two days later, the shuttle docked, Andy and Dave changed places, and we went through our almost routine task of supporting a space station with the shuttle. The technical details were nearly perfect, but we were soon confronted with a human issue.

During the crew changeover, each new Mir crewmember would install his equipment in the Soyuz to be prepared to leave quickly in an emergency. This included donning and pressurizing the Sokol space suit to ensure it was airtight. The Sokol suit, worn only during launch and entry in the Soyuz, is a very basic pressure suit designed to keep the crew alive and functioning in the event of a loss of pressure. It is made to fit as tightly as possible, with laces to adjust the final fit to the individual. The fitting had been done in Russia long before Andy had launched on the shuttle, and this was his next opportunity to see it.

Andy could not don his suit. It was more than just tight—it did not fit. It is well-known that an astronaut's spine expands after being out of Earth's gravity for several days, and an allowance for this growth was supposed to have been built into the suit's fit. For some reason, this time, it was not sufficient.

The support team instructed Andy on how to readjust the fit. But while he followed their instructions, Victor Blagov jokingly told a reporter that Andy was not the kind of tough man the Russians would send to space. We were dumbfounded by Victor's gratuitously antagonistic comment, especially since the issue was apparently caused by the actions of a Russian suit technician. When Andy heard about it during a news conference, he handled it well, but he was deeply upset by this unprofessional dig. I told Scott to reassure Andy that I would follow up with Blagov later rather than draw attention and make the situation worse.

Then, immediately after the shuttle undocked, Solovyev decided to stir things up by objecting to the presence of a bioreactor experiment on the station, claiming it was unsafe. Despite its many months of safety evaluations, he threatened to throw the experiment into the Progress to burn up with the trash. Since this was Andy's prime experiment, he was extremely upset. Despite performing perfectly, he was the subject of two confrontations in the opening days of his mission. It was a severe test of his mettle and one that he passed with flying colors.

Frank called me to share this news while I was driving home after the shuttle undocking. I pulled into a Home Depot parking lot while the team assembled, and within minutes, we had Scott Gahring, Victor Blagov, Rick Nygren, and others on the line. Rick Nygren confirmed that the approvals were all in place and that the experiment met every safety requirement.

For some reason, Blagov was enjoying our discomfort at the threat to our main experiment. This made Frank angry, and he reminded Victor that we were critical to keeping Russia's space program alive. Eventually, Victor agreed that all was in order and would tell Solovyev to leave the experiment alone. Once again, I wondered how we could trust people who behaved so childishly.

The mission settled into a fairly smooth tempo. When the Russian crew changed out a few weeks later, Andy finally had the crew with whom he had trained. Still, one more big problem awaited the Phase 1 team—this one completely of our own making.

Throughout the program, I listened to the air-to-ground communications with the Mir as often as I could. Most of their limited time was filled with Russian conversations about the station systems, with only a few moments for NASA's use.

One day, I heard Andy anxiously telling Scott that the routine task of replacing the nutrient fluid in the bioreactor had failed. Instead of filling the vessel with fresh fluid, the procedure had sucked it out.

"The cells are floating in midair in the middle of the chamber," he reported. The exasperation was plain in his voice.

Scott told him to stand by while the ground team looked for a solution. They both knew that the next comm pass was the last of the day, and we needed a solution before then.

While Scott was still talking to Andy, I called Dr. Neal Pellis, the principal investigator for the experiment. I told him what had happened and to expect a telecon from his ops team within moments.

"You have ninety minutes to save your experiment," I said. "Use them well."

With our team spread out across the planet, I had arranged for a "meet me" telecon line that could be used for exactly this kind of urgent matter. I dialed in to hear how the team was performing. Within moments, I heard a dozen anxious voices.

Scott and the science lead took control of the meeting, recounting the conversation with Andy and the timeline for getting an answer to the problem. Another voice reviewed the procedure that Andy had used and how it was supposed to refresh the nutrient fluid. They quickly went around the loop to get everyone's view of the problem. Opinions flowed freely. After an hour, I broke in to remind people of the time remaining. When the time was down to five minutes, I interjected again.

"It is time to decide who is going to talk to Andy and what that person is going to say."

The final message was reviewed, and the line went dead as Scott prepared to give Andy the procedure. When the comm pass began, Scott passed on the new procedure, and Andy repeated it back to make sure that he understood what he was to do. Everyone knew that this was the last chance to save the most important research of his time on Mir.

Then, they signed off. We would not know whether the procedure had worked for at least ten hours. It was a very suspenseful period for the science team, with lots of time to reflect on how they could do better in the future.

When Andy called Scott the next morning, he was jubilant. The procedure had worked perfectly, and the bioreactor was functioning smoothly.

21. TRANSITION TO ISS

Shuttle flight STS-91, which would bring Andy home, was the final shuttle mission to the Mir, and it occasioned strong feelings across the team. Although I felt sad and nostalgic as this intense experience drew to a close, I looked forward to sleeping through the night without worrying about waking up to some terrible surprise.

The manifest came together remarkably easily. Because we were not leaving a crew, we did not have to worry about sending up food, clothing, and science materials, nor did we have the innumerable little training and planning crises that had been part of every other crew's mission. The shuttle even had room for a precious science payload, the Alpha Magnetic Spectrometer (AMS). It was almost like a "normal" shuttle mission where NASA was in control of everything.

Another change was that Valery Ryumin, the Russian cosmonaut and program manager for the Mir, was going to fly on the shuttle. With so many problems in the aging station, Ryumin wanted to see its condition for himself to judge its future prospects. Getting agreement to let him fly had been difficult because his English proficiency was not good, he smoked, and he was well above the weight our doctors thought healthy. Despite his promises to address these concerns, none had improved much as the launch date approached.

He also continued to manage the Mir program, which took a lot of his time. Although he remained in Houston, he spent hours each day on the telephone to follow what was going on. This came at the expense of his shuttle training, which upset his NASA colleagues. As an experienced cosmonaut, he apparently felt he could handle anything that might require his attention. He also saw himself as a senior executive—a passenger to be supported, not a working crewmember. However, shuttle missions are extremely busy, and Commander Charlie Precourt did his best to get Ryumin to do his part. Many people grew angry that Ryumin's misbehavior was tolerated, and it soured his relationship with some who had supported the Phase 1 program. Even Frank Culbertson's irritation sometimes showed.

Despite our preflight misgivings, the mission itself went swimmingly. Andy was happy to come home, the AMS collected some interesting data, Rick Nygren and his folks wrapped up their research program, and we were thrilled to bring this stressful time to a successful close.

For me personally, the program's end left my future uncertain. Because Phase 1 was organizationally part of the ISS program, we all assumed that we would be absorbed by that organization. The timing was good because the ISS deputy for operations, astronaut Kevin Chilton—known to all as Chilli—was going to return to active duty in the air force, allowing us to fill his slot

Final Accounting

As we were wrapping up the program, the property management organization sought to account for the materials we had purchased. Many things, like food and clothing, were understood to be disposable. Others, like laptop computers and experiments, were both valuable and easily stolen, so they had to be accounted for. With just one shuttle to bring everything home, much of that equipment was left on the Mir. Needing a final disposition for these items, I told the property managers that they would burn up when the Mir eventually returned to Earth, but they had no way to list that in their system. Meanwhile, some crew tried to return unauthorized mementos, what Tommy Holloway called "contraband," that could be sold on eBay or through auction houses. I had to manage both.

while bringing our experience and relationships into the larger program. Everyone assumed Frank Culbertson would slip into Chilli's job, but because Chilli had no deputy, it was not clear what role I might have.

As the mission wound down, Frank came to my office.

"I want you to know that I appreciate how selflessly you have dedicated yourself to supporting me," he said. "You have gone far beyond the role of deputy. Many times, you have had to carry the full burden of the program when I wasn't around, and you did so extremely well. You've never gotten the attention you deserve, and I plan to correct that. You and I will work on everything together, far more as equals, even as codirectors."

His assurance comforted me. Although I was proud to have served as his deputy, there were many times when he'd taken my participation for granted, often referring to me as "staff." Despite my humble start, I felt I had earned my spurs as well as he had, and I dearly hoped his words were genuine.

When we were finally free of the threat of another crisis, each of us scheduled several weeks of vacation. Frank left first, and a few weeks later, I took a luxurious three weeks in the Adirondack Mountains. It was almost six weeks before we saw each other again.

When we did, I hardly recognized the man I thought I knew so well. He practically ignored me, breezing through the office and giving directions as if he were the only Phase 1 veteran. I assumed that this was prompted by his relationship with Randy Brinkley, who remained the ISS program manager. Randy preferred to make decisions without team input and pressed people to do things his way. This felt like a repudiation of everything we had learned in the Shuttle-Mir/Phase 1 program, where our success was the product of a superb team effort. Many times, when members of the team disagreed with our decisions, they set aside their doubts because they knew their inputs were heard and valued. That sense of trusting teamwork was fading under Randy's hegemony.

As I waded back into the ISS program, my experiences during Phase 1 left me extremely concerned about its future. I believed that Frank and I had a duty to use our experience for the benefit of ISS. Although Frank obviously agreed, I felt he was going about it the wrong way.

My most immediate concern was how the operations team would be led and by whom. The Russian operations team for ISS was largely the same as for Mir, and while they had worked well with their MOD colleagues during the shuttle visits, they were deeply skeptical of following our ISS leadership. This was a critical problem; many ISS operations would be performed by Russians and their hardware, and NASA needed a strong, respectful relationship with them.

Ryumin seemed particularly upset by the changes in our relationship. Part of his anger stemmed from the antagonism surrounding STS-91, but he was also distressed that ISS leadership had paid so little attention to Mir. While Ryumin would have a lesser role in ISS, he was the Russian who knew NASA best, and his opinion mattered greatly. When he was angry, it hurt us.

Some NASA colleagues were also a problem. They considered our Phase 1 work to be a minor sideshow compared to their "main event" and felt free to ignore its lessons. I was almost apoplectic. We had just spent three agonizing years building a critical relationship and learning priceless lessons for ISS, only to have the program seniors disown our efforts.

As had happened with Boeing a few years before, serious tensions were developing between NASA and its partners, especially the Russians. The most urgent problem I saw was the huge disconnect regarding roles and responsibilities. The ISS program office had put together a large set of planning documents describing how they wanted things to run, but their NASA-centric approach had little support from the partner nations who did not want to be stuck with whatever NASA dictated. Our Canadian friends seemed willing to look the other way, but both the Europeans and the Japanese were more circumspect, seeking to give more input and receive less direction. Yet, whenever I brought up the need for a carefully crafted set of roles and responsibilities with Frank or Randy, I was told there was no issue. In Randy's mind, NASA was totally in charge, and the internationals would behave as if they were contractors. I believed Frank shared my concern, but he wouldn't challenge Randy so early in their relationship. His silence bothered me deeply as I knew time was running out.

Once again, I saw disaster ahead. I was certain the Russians would not accept the role NASA's ISS team had designated for them, yet nobody on the NASA side was willing to address the issue. I started to avoid Frank to keep from showing my anger.

As the summer wound down and Phase 1 people began moving to new jobs, celebrations of our success were planned. As the day of the biggest party approached, I was upset enough that I considered not going to avoid making a scene with Frank.

"Oh, you two little boys need to grow up!" scolded Jessie Gilmore, Frank's longtime secretary. "The party isn't for you—it's for the team! You owe it to the people who worked so hard for you these past three years to give them a night to remember. You need to get yourself to the party and act like you're enjoying yourself!"

She was right, of course, and as the night approached, I did my best to get into a positive frame of mind. It turned out to be easier than I thought, as the energy and enthusiasm from the team were palpable. When Doris and I arrived at the party, we found a room full of happy, welcoming people savoring the moment. I was glad Jessie had scolded me into coming.

Soon after entering, an unfamiliar man raced across the room and extended his hand in an enthusiastic greeting.

"Jim, Andy Chaikin," he blurted. "I am so glad I finally get to meet you. I've been such a fan of your work!"

I was both startled and flattered. I knew of Andy from his extensive writing on the Apollo program and was surprised that someone who knew every major player in Apollo would be eager to meet me!

His boyish enthusiasm immediately lifted my spirits and made my evening truly memorable. We remain friends to this day.

The Phase 1 team enjoyed themselves thoroughly, and I was soon happily sharing stories and laughing heartily. Although our team had worked together closely for over three years, only at the end did I truly appreciate the camaraderie that bound us. I vowed to pay more attention to my ISS colleagues as we worked together in the years ahead.

Soon after the party, as its happy glow was beginning to fade, I learned that another writer, Bryan Burrough, was preparing to release a book about the Phase 1 program—one that had the potential to be either very good or very bad for NASA. When he'd first approached the agency about it a year earlier, our public affairs lead, Rob Navias, had come to see me.

"Bryan is a big-name writer, and this is going to be a major book. But he writes sensational stories, and we need to make sure he gets the facts right. We need to manage his access, not just let it happen."

"How would you propose we do that?" I asked.

"Either you or Frank should take the lead in explaining what's been going on and why. You need to teach him about the technical work, let him see the challenges of international cooperation, and make him understand that this is an important program operating on the frontier."

At the time, we were in the middle of our recovery from the collision, and both Frank and I were overwhelmed with work. Since Frank was so often called away, I had little choice but to take on the role.

Bryan came to see me a few weeks later. Personable and intelligent, he showed great enthusiasm for his new task. He also freely admitted that he knew nothing about space and would need a lot of help with the subtleties.

"I'll do my best to help because it's important that you get this complicated story right," I said. "Just please keep me in the loop on your work."

"Absolutely! I don't want to make a mess of this any more than you do. We'll go through the draft together before I send it to the publisher."

That seemed like as good a deal as we could have expected.

After that, Bryan called often and even came by my house to discuss important sections of the book. He seemed anxious to make sure he correctly understood the events and their significance. His interpretation was usually pretty close, but there were things he'd taken completely out of context. He was a quick learner and generally accepted my corrections. I thought he'd been on a constructive track.

Nearly a year of intensive work passed, and the book was coming together. Then, after the party he called to say he had to break his promise to let me read the manuscript.

"The publisher wants to get the book out for the Christmas season, and I'm busting my butt to get it to him."

This was alarming but out of my hands.

Shortly before the Christmas holidays, he sent me an early copy of the book, titled *Dragonfly* for the Mir's appearance, before it hit the stores. When I scanned its pages, my heart sank. Not only was it written in the breathless style of someone sharing dark secrets, but it also contained countless quotes that I knew had been created from whole cloth. Some were just embarrassing, while others had the potential to cause great harm.

I quickly jumped ahead to see what he had written about certain events that could easily be misunderstood. Almost immediately, I was shocked to find a passage that could do great damage to someone who deserved only praise.

Dr. Al Holland was the psychologist who worked diligently to help us care for the emotional health and well-being of our crews. He worked quietly in the background with his Russian colleagues to assess the emotional and psychological environment on board. After each mission, he conducted detailed debriefs with the crews to evaluate what could have been improved before offering his best advice. His role was sensitive, and his conclusions were kept highly confidential. We all worked to protect our crews' medical privacy.

Astronauts have always been skeptical of doctors, especially psychologists and psychiatrists, and often view them as people who could ground them over a simple misunderstanding. For this reason, Al was extremely careful to ensure he did nothing to violate the trust he had earned within the community. In his book, Bryan quoted Al as describing how John Blaha had become depressed during his time on Mir. Although John had publicly admitted this, including to the media, Al would never have discussed such private information with outsiders. Al only told Frank and me what we absolutely needed to know; we were all bound to strict rules of confidentiality.

I called Bryan immediately to ask why he had put John's words in Al's mouth.

"Heck, John told me all about it. What's the big deal?" he asked.

"John can do that if he wants," I said, "but Al cannot. Such information is protected by a strict code of professional ethics. Since John told you himself, why did you attribute the quotes to Al?"

"I just thought it sounded better that way," Bryan replied.

Bryan's breezy decision had the potential to destroy Al's career and undermine a critical component of the astronaut support system.

Feeling responsible, I set out to correct the misunderstanding, immediately visiting George Abbey and the leaders in medical operations and flight crew operations. Repeating Bryan's explanation, I pleaded with them to help me preempt the worst

Creative Nonfiction

It is common practice to include dialogue when writing nonfiction, even when no transcript of the conversation exists. In this work, I have only included conversations in which I was a principal participant. Although the wording may not be exactly accurate, I have done my best and used my extensive contemporaneous notes to ensure that the meaning and intent of the conversation was preserved.

of the damage the incident could have on Al's reputation and work. They all agreed to do what they could.

Then I called Al. He had already seen the quote and doubted the damage could be repaired. But as it turned out, his reputation as a fair and honest professional was more powerful than that one fabricated quote.

Worse trouble arose from the book's plentiful dubious information, often contained in dialogue that Bryan could not possibly have heard from the principals. The whole reason I had spent so much time with him was to prevent this kind of speculation. I had even given him access to some transcripts of particularly sensitive events, like the crew reports of the collision. But the quotes that Bryan had invented often left a bigger impression than the transcripts of actual events.

Far and away, the most galling aspect of the book was his lionization of Blaine Hammond. This notoriously untrustworthy astronaut had made no effort to express his concerns or even check his facts with us before sending his ignorant and inflammatory letter to the IG. Yet, he had somehow convinced Bryan to cast him in the role of the lone voice of sober reason—the courageous truth teller in the midst of sycophantic bureaucrats. Nothing could have been further from the truth.

It did not take long for the negative reviews to hit home. The folks in Phase 1 knew that I had spent a lot of time with Bryan, and the magnitude of my failure was obvious to them. Others, including some who should have known better, took what they liked from the book and ignored its obvious errors. This was especially true for those who wanted to bash the Russians for their unsophisticated ways or who were opposed to their inclusion in the first place.

Fortunately for us, the book had far less impact than we feared. The very style of its storytelling helped many see that it was written to be sensationalistic rather than factual.

The book's most serious consequences appeared to be to my career. Frank was obviously hurt and angry, and although we never discussed it, I knew he felt betrayed. If the book had come out before STS-91, when Frank and I had still been working in a tight partnership, I am sure we would have aired it all out and gotten past it. But because we had become so distant over the summer, we never did, and I felt the lingering aftereffects for many years.

I found a little relief in an anecdote my wife related. She had encouraged a technical contact of hers, an engineering professor at Texas A&M, to read *Dragonfly*, telling him her husband was in it. My wife had not taken my last name, so when the professor had finished the book, he'd

> ### Bryan's Email to Me
> I hope you know I'm sitting here on pins and needles as you read this book. Someone much smarter than me once said that every author writes their book for one person, and in this case, I feel like that person was you. No doubt we will differ on some of my characterizations and opinions. But whatever our differences may ultimately be, please don't ever think that I have anything less than total respect for what you, Frank, and the rest of the team accomplished here.

asked, "What was your husband's name again?"

"Jim Van Laak," she'd said.

"Oh! He was my hero. He was the only one who seemed to know what was going on."

It was nice to know that at least one person thought I had survived the experience, but it was a small comfort at the time.

The controversy over *Dragonfly* was one more reason to throw myself into my new job, which gave me more important things to think about. One of the most urgent was resolving the roles and responsibilities for Frank and me. Without an explicit agreement, he focused on communicating with senior management and the media while I dealt with the day-to-day operations. I gave it all the energy I could muster, though compared to the urgent crises on Mir, the workload was almost leisurely.

We made steady progress with the work, but we were no longer the close team we had once been. Neither of us wanted to have the difficult conversation that awaited us, so we just kept working on what was in front of us.

22. MOVING ON

ISS had no immunity from the problems that come with any new, large, and complicated development program. Hardware delivery delays continued in both the US and Russian segments but for different reasons. The Russians faced few technical issues since their designs had flown before, but they fell behind schedule due to financial problems. Meanwhile, NASA had sufficient funding but struggled with new designs and the challenge of integrating the program's many elements.

The first two modules to be deployed—Russia's *Zarya* module and the US *Unity* node—were ready for launch eighteen months before the ones that would follow them. We had to decide whether to launch them as soon as possible or wait until the rest of the station was ready. I argued for a delay, concerned about the reliability of these two elements without the rest of the station to share the load. Others argued that it was imperative to launch as soon as possible to keep the workforce engaged and mollify our critics. Ultimately, our leadership chose the early launch option.

Throughout the arguments about when to launch, nobody had foreseen the best reason to proceed with the launches: forcing the new operations team to start working together while their responsibilities were modest. Although I did not appreciate how far behind we were, I could see that the team was not yet ready for the main event. While they managed well for the first two launch missions, it quickly became evident that they needed every bit of flight experience possible before tackling the major challenges ahead. I was never happier to have lost an argument.

ISS operations began on November 20, 1998, with the launch of the first Russian-built module from the Baikonur Cosmodrome. This was the element that had been the key to bringing Russia into the ISS program. The newly christened *Zarya*—Russian for "dawn"—was followed two weeks later by the first American element, the *Unity* module, which was launched aboard the space shuttle. They docked in December 1998 with great fanfare. Both missions were executed superbly, and the public reaction was very positive.

We had expected the eighteen months until the next element launch to be slow and tedious, but instead, we found it to be a challenging learning experience. Both the NASA and Russian teams were ready to support their own elements, but neither was truly ready to execute a joint program. While the hardware worked together, the flight control teams did not. My concerns about the reliability of the early hardware were addressed by putting the equipment into a dormant state and activating it periodically only for minor orbit adjustments and maintenance.

Although there was plenty of blame on both sides, NASA, as the program leader, bore the primary responsibility for this organizational failure. Despite the fanfare of our international partnership, few Americans made the effort to understand or respect the Russians—and the Russians reciprocated. The plan to treat the Russians as if they were contractors quickly hit a brick wall.

The principal NASA links connecting Phase 1 and ISS were Frank Culbertson, me, and a small group who had worked in Moscow. With Frank campaigning to become the next ISS program manager, much of the responsibility for integrating Phase 1 experiences into daily practice fell on the ops leads and me. The ops leads did their best to share lessons learned, but they had little influence. I had more clout, but resistance to change was fierce. People who should have been eager to learn from our experience instead fought against us, wasting time and draining energy that was urgently needed for future preparations.

There were many excuses for their resistance. One group groused about having "communists" involved. They would not accept the end of the Cold War or trust our former adversaries. Most of these complainers did what they were told half-heartedly; although they rarely blocked our work, their tepid behavior stood out in an agency filled with over-performers. Others acted from personal and organizational parochialism. They fought to protect their personal positions without concern for the impact on the program. Some still resented the cancellation of the Space Station Freedom program. Many people had lost their jobs when that program had been canceled, and while some had found new jobs in ISS, many had not. In fact, I was the only senior manager from Freedom to go directly into ISS.

The most common reason for this resistance was that many people just did not accept the need for change. They thought NASA led the world in space and had nothing to learn from others. They dismissed lessons learned from Phase 1 with "But this is not Mir," refusing to even consider applying its lessons. They saw a contest between "our way" and "their way." Sadly, their childishness led them to ignore the Americans who had learned so much from Phase 1, including myself.

The Russians were not much better. Their space program holds a special place in their culture, and they found being forced to partner with us deeply humiliating. Their leadership did their best to preserve as much of their program's independence as possible and planned to operate their segment separately—just as Mir had functioned independently when docked with the space shuttle.

As I had worked to heal the rift between NASA and Boeing, I now worked to bring the Russian and NASA teams together. I wanted to ensure that the next NASA operations manager—whomever they turned out to be—had the authority needed to manage the effort. I knew this process would take months to accomplish, but the extended schedule gave us the necessary time before full-scale assembly began.

Three additional space shuttle missions were added for maintenance, logistics delivery, and team training, providing us with invaluable experience. Once again, circumstances created the perfect opportunity for us to learn what we needed.

Challenges soon arose, and one of the most painful occurred during the first interim shuttle flight. During one Mission Management Team (MMT) meeting, a disagreement emerged over how to configure a pressure equalization valve on the *Zarya* module. The Russians wanted to set it in a position that allowed remote operation in case a Russian Soyuz had to dock before the next shuttle mission. Because the action would slightly increase the risk of a leak depressurizing the module, Jay Greene, a senior ISS manager, objected.

Jay was one of many Apollo veterans who had gone on to become senior leaders at JSC. A member of Gene Kranz's famous "White Team," which had led the first moon landing on Apollo 11, Jay had later been a flight director and held other management roles. Although he was extremely sharp technically, his personal arrogance, explosive temper, and caustic behavior often did great harm.

Technical disagreements are normal; what matters is how they are presented and resolved. In this case, Jay had a valid technical point, but the risk was very small. Because the Russians had designed and built many elements similar to this one, they were confident in their approach. Jay, however, knew nothing about the hardware beyond the brief summary presented in the meeting. Had the decision been mine, I would have accepted the Russian recommendation as a gesture of trust and respect for their expertise. Instead, Jay—who despised the Russians—escalated this minor issue into a major conflict. In front of the entire MMT, he denounced the Russian proposal as idiotic. Frank Culbertson, who chaired the meeting, was unsure how to proceed. With time running out, he sided with Jay, causing damage that would haunt the program for years.

Shortly after the shuttle flight ended, key management changes were announced. Randy Brinkley was leaving NASA for a job in industry and would be replaced by Tommy Holloway from the Space Shuttle program. Frank Culbertson, having lost his bid for program manager, was pursuing a flight assignment. This made me the acting head of operations and set up an immediate conflict with Holloway.

After Holloway left Phase 1 to head the Space Shuttle program, he paid little attention to the ISS program's upcoming needs or the consequences of his decisions as manager of Shuttle-Mir. As a result, he was woefully unprepared for his new job. Worse, he had a terrible relationship with Russia. For reasons I never understood, he was also very hostile toward me. He told me that my role in the program was far from decided and that he strongly disagreed with my ideas about how to work with Russia.

This began one of the most curious and disturbing periods of my professional life. For almost a month, Holloway intentionally sowed conflict within his entire management team, apparently in search of any dirt that might exist. Calling individual managers into his office, he would repeat—or possibly fabricate—alleged insults and innuendos from others. In my case, he focused on Jay Greene, with whom I had long had a difficult relationship, but others told of similar experiences. Whatever he thought he could gain through this behavior, it certainly undermined the trust and cohesion of his team.

He could not have done this at a worse time. The ISS faced great challenges in those early days because our small station was fragile and our partnership was weak. Instead of building confidence, teamwork, and flexibility for dealing with the obstacles ahead, Holloway chose to tear down his predecessors, attack his subordinates, and plant fear and anger throughout the team. While this solidified his power, it undermined every other aspect of our work.

Our arguments were particularly intense. When speaking with me, he often pounded the table, insisting that NASA could not be responsible for the future of the station because "the Russians are going to do whatever they want to." Despite the enormous trauma of our Phase 1 experience, he was willing to accept the same weak posture he had taken during Shuttle-Mir. His position reflected his failed experience as manager of Shuttle-Mir, and while that might have been excusable when we were guests on the Russian space station, it was completely inappropriate now that we were responsible for this $100 billion program. It also violated every treaty we had with our partners and the promises we had made to Congress and the White House. When I pointed this out, he ridiculed me as naïve and called those promises foolish.

One of my strengths is that I listen to everyone, as I did with his comments, even when he attacked me. I kept an open mind, considering whether he might know more about some aspects than I did. However, his actions solidified my views against his. Again and again, he displayed blithe ignorance of the challenges we had overcome to develop the ISS and the problems we had survived—many of which he had helped cause—during Phase 1. He proudly claimed to know little and not care about them, including denying any responsibility for the problems on Mir.

This created a personal and ethical crisis for me. Should I take direction from my boss or stand up for what I knew to be true? Should I be humble and accede to his seniority or boldly stand up for what I felt was right? It was another life-defining moment.

In the end, there was only one possible decision: The future of NASA's human space program was on the line, and I had to act on the history I had seen and the truth I knew. When we talked, I held nothing back, politely but firmly telling him that his views were simply wrong and that I would not follow his direction.

"You get to choose your head of operations," I said, "but if you choose me, I will use every bit of experience and judgment I have to do what I think is right!"

Holloway was angered by my response and scolded me as if I were a child. He ridiculed my views and told me he knew far more about this business than I could imagine. He did his best to intimidate me, perhaps to drive me away, but I surprised myself by standing my ground. Although I desperately wanted to work on ISS, I knew how close Phase 1 had come to disaster and would not willingly go down that path again. He'd made it my duty to oppose him when I thought he was wrong, no matter the cost, and I lived up to that duty. Certain I had lost the operations manager job, I resigned myself to another career setback. Worse, I envisioned the precious ISS partnership breaking apart.

A week later, Holloway shocked me by asking me to take the operations job—but with a major condition. He said Jay Greene had threatened to resign if I was made his equal as the deputy for operations, the position that Kevin Chilton and Frank had held. Tommy gave me full responsibility and authority in the deputy role but with a demotion to the head of a newly created Mission Integration and Operations Office.

I was furious at Holloway for caving to Jay's demands, especially given the enormous responsibility he was asking me to assume. As chair of the MMT, I would be responsible for the safety and productivity of the most fragile and expensive vehicle in the world. In the mission integration role, I would be responsible for the design and execution of the many complex missions to come. As head of the operations team, I had to build and lead our international team through whatever crises that might arise. At a time when he should have been lifting the status of my office and certifying his confidence in my leadership, he was demoting me and leaving a visible void on the organizational chart. I was heartsick.

Holloway's claim that I had the same authority but in a different box on the organizational chart ignored the effects of the reduced status of this critical role and the lower grades it would impose on my division managers. He weakened the most vital part of the program at its most critical moment.

Although deeply angry and embarrassed, I accepted the job because it appeared to be the only way to bring Phase 1 experience into ISS leadership. It looked like the beginning of another crushing job with the potential for a disastrous outcome for which I felt enormous responsibility.

A few days later, one of my interpreter friends told me that Ryumin had told Holloway that I was the only American the Russian team trusted to lead operations. I was stunned. My relationship with Ryumin had always been strained, and I had struggled to establish NASA's value with our Russian colleagues. Upon reflection, I supposed that after all we had been through together, Valery was confident that I would listen to their views and do my best to make balanced decisions.

In the months since the end of Phase 1, I had devoted my efforts to giving the NASA operations leader—whomever they turned out to be—the chance to succeed. Suddenly, it was very personal.

23. BUILDING THE ISS TEAM

As I drove home after accepting the job, a familiar thought kept running through my head: "Be careful what you wish for because you just might get it." Part of me wanted to laugh, and another part wanted to cry. Although I had been hoping for the job of operations manager, I never thought I would actually get it. And now that I had, I was crippled by an antagonistic, unsupportive boss and terrified by the responsibility and challenges that lay ahead.

My first priority—making this grand partnership work as a high-performing team—cast an intimidating shadow over my future. Opinions on how to achieve this were as numerous as the people involved, and my boss had absolutely nothing constructive to offer.

The consequences of Tommy's decision to demote my position rolled through the organization. The Mission Integration Office led by the very capable Keith Reilly was merged with the Mission Operations Office run by Chirold Epp. This instantly demoted two outstanding division chiefs who had done nothing but excel in their jobs. As a result, Keith quickly chose to leave, and Randy Brinkley welcomed him into the industry. Fortunately, Susan Creasy stepped into his role without missing a beat. Chirold, another of the original ISS managers hired by Shep, was clearly disappointed but decided to stay to see the culmination of his years of work.

While I was angry about the new organization, it was easy to be enthusiastic about our work. We were leading the construction of this fantastic new vehicle and had the best seat in the house. The most intensive period of human spaceflight operations in history was rapidly approaching, and we were at the center of it all. The launch package managers led planning for the assembly flights, while increment managers did the same for station operations. Others on our team managed the operations contracts, ground facilities, logistics, and training.

The workload was incredible, but rather than wearing people down, it energized them. Every day was filled with challenging, exciting work, and the team loved it. Deadlines were real and looming, requiring important decisions to be made throughout the organization. For most of the team, morale was at an all-time high.

My biggest challenge was to build one cohesive operations team using experts from all our partners, with an emphasis on the US and Russia. Both of these proud communities saw themselves as the world's expert, and neither was willing to play second fiddle to the other. As an American, I was sympathetic to NASA's approach and proud of its accomplishments, but my Phase 1 experience had shown me similar strengths and weaknesses in the Russian approach. My goal was to maximize our collective strengths and minimize our weaknesses.

Despite my office being lowered on the organizational chart, my authority and responsibility remained undiminished. My role as MMT chair kept me at the top of the decision-making process, though I had a nagging fear that the position had been weakened in the eyes of others. Holloway remained convinced that the Russians were uncontrollable and adamantly refused to back me up when I made it clear that they would have to follow program leadership. Going into the defining role of my career without leadership support, I had no choice but to minimize direct confrontations.

To avoid conflict during normal operations, I began mapping out roles and responsibilities across the program—including those of our international partners—just as I had with Boeing. The documents NASA had prepared were a good start, but they reflected its parochial perspective. Their tone was directive, leaving little room for alternatives, which could discourage cooperation. Phase 1 had taught me that all parties needed to have a voice in decision-making, and I was determined to make that happen.

Thanksgiving vacation gave me some time and distance for creative thinking. With things quiet on our little station, I flew the family to Michigan for the long weekend. While my wife and kids enjoyed time with my in-laws, I got to work.

Starting with the NASA documents, I crafted a paper that broke our work into functional parts. It applied systems engineering principles to define the functional elements of the program organization and their mutual interdependence. My goal was to capture the essence of what each organization needed to do—including their inputs and outputs—as a foundation for negotiations. The description didn't need to be perfect; in fact, having the participants correct a few errors would increase their ownership of the final product. It would also give them the opportunity to suggest modifications to NASA's plan.

After the holiday, I shared my work with the team. My first stop was with my deputies, Susan Creasy and Chirold Epp. They agreed with the content but felt it merely repeated what was already in the program documents. While they were right on one level, they didn't see the difference in how I had framed the work. I needed to gain support from other teams and work through the many open issues that remained.

My next stop was the Mission Operations Directorate (MOD), led by my friend and confidant Randy Stone. Randy patiently worked through the issues with me, correcting numerous errors and anticipating future problems. Many of the boxes on the chart represented functions within MOD, with counterparts in Russia, and Randy wanted to know how I planned to coordinate their efforts. Together, we developed a model for where coordination was required, and I explained my expectations for how the partners should collaborate. The rest of the details would be left to the experts, with me stepping in only if they couldn't reach an agreement. Randy liked what he heard, but I sensed something deeper was on his mind.

The next day, he came by my office.

"Our conversation got me thinking," he began. "Moving into the station era isn't just the next chapter for MOD—it's a profound change. Up until yesterday, I thought it was our normal business being paid for by a new customer, but now I see it's much more than that."

"Please go on," I said, eager to learn from his insight.

"Apollo was a program, and all its missions were linked by the same basic goal. But the missions were more or less independent of each other, and the hardware wasn't reused. If one failed for whatever reason, we could simply repeat the mission with minimal delay. The same is true for the Space Shuttle program, since it serves many customers with unique missions. But with ISS, each mission builds on what came before and supports what follows. Each one is intimately tied to the rest and fits in a unique place in the sequence. If we have a failure, it could severely impact the rest."

"Which is why I've asked for alternative assembly sequences," I said. "The probability of suffering some kind of a setback over five years and dozens of missions is significant. We need to be ready to respond."

Randy nodded before continuing.

"I saw that right away and applaud your efforts, but our conversation yesterday made me realize how this also changes the way MOD does its job. During Apollo, each mission was like a scene in a play—part of a narrative, but each with its own characters and subplots. Then, with the Space Shuttle program, each mission became a little play of its own with even more independence. Individual flight directors had lots of room to shape their work to fit their personalities. But now the station is changing that. Now our missions are acts within an enormous, complex play. We have to meet the station's needs within its narrative, not ours."

"Interesting point! How hard is that going to be?" I asked quietly.

Randy shook his head somberly.

"I don't know. The organization is talented and flexible; it can do whatever the job requires. But people can get set in their ways, and it won't be an easy change for the old guard. Some of them may not make it."

After my conversations with Randy, I called Russia to share my work with Victor Blagov, their senior flight director. Knowing he had little patience for elaborate plans, I kept our conversation focused on how the flight control teams would have to work together, and he was not impressed. He did not like all the coordination I described.

"We should operate like we did during Phase 1," he said through the interpreter. "You can run your side, and we will run ours. We don't need so much talking."

"Victor, ISS is nothing like Mir! We do not have two independent spaceships—we only have one. This station is one machine with one brain. We need your power, cooling, life support, and more, and you will need ours, especially in the early days. This is what our leaders asked for, and it is what we designed!"

He was not giving up.

"Jim, we can do that like we did on Mir with a joint plan for just the actions you need to take. When you need a special attitude, we can provide it. If you need to borrow some power, we can give it. But most of the time, we want to be on our own."

"Victor, I helped design this machine, and that will not work. Our leaders insisted that we make one joint station, and that is what we did. Now we need to work together like crewmates on the same ship—because that is what we are."

I could tell that he was still unhappy.

"How can we do this when NASA does not know our systems and operations? Look at what happened with the pressure equalization valve!"

"You're right about that, but now I am in charge of operations, and Jay Greene will not interfere. I promise!"

"But it is not just Jay Greene who does not listen! None of your leaders understand how our systems are designed or how they are used. They do not have our experience of keeping crews in space for many years." The exasperation in his voice was genuine.

"You're right about that, too, but I am working on it as hard as I can," I said carefully. "I have been teaching my colleagues about your history, about your systems, and about how you operate them. Please work with me so that I can show them that we can be good partners in running our new station."

Victor could see my honest efforts to bring balance to the relationship, and as we talked, he began to see how my plan might have value. Although he still thought his model was fine, he agreed that my approach was acceptable. I counted this a huge victory despite the many issues that remained.

The crew office was generally pleased with my approach and added several issues I had not tried to address. They were particularly concerned about the Russian practice of paying cosmonauts to perform certain tasks, like EVAs, and worried that it could incentivize them to do things that could cause problems. They were also concerned that it could create tensions that might divide the team. Our experience with Mir supported that concern, but there was nothing I could do about it in the near term.

The ISS engineering team was less supportive. Although I had an excellent working relationship with the line managers, Jay Greene's contempt for me and everything Russian undermined some important NASA work. The scope of the problem became clear when I sat in on one of Jay's design reviews and noticed that there was no mention of the Russian systems that were connected to ours.

After listening to the discussion for perhaps twenty minutes, I had to ask the unspoken question.

"I'm just curious. You have a great description of the US segment, but how do you account for the interactions with the Russian systems? I don't even see the interfaces here." I avoided looking at Jay when I asked my question, but I could sense the rage building at his end of the table.

Silence filled the room, and nobody moved a muscle. Even Jay was silent. After a long, pregnant pause, my old friend Barry Boswell quietly spoke.

"We don't talk about them here."

After a few more seconds of awkward silence, the meeting continued, team members discussing the American elements as if I had never spoken. Although not completely surprised, I was horrified to see the ISS team ignoring what was clearly their responsibility. I needed to make sure that somebody was doing this critical systems integration work.

After leaving the meeting, I asked Boeing's program manager, Doug Stone, what he knew of this.

"Jay refuses to touch anything Russian," he said bluntly. "He is only reviewing the American segment. Boeing has the responsibility for system-level integration. We think we have it under control."

While this made sense, and I had confidence in Boeing, it still bothered me—both because Boeing was not party to the treaties between nations and because I knew NASA could not escape its responsibility. I next went to see Randy Stone for his take on it.

"No question that Jay dropped the ball on this, but I think the rest of us have it under control," he said. "Our MOD systems experts are doing what they do best. As you know, they build detailed system drawings and procedures that include everything in the system to make it all work together. Boeing is eager for our input, and we are working well together. Like we talked about the other day, we need help from the Russians, but we will be fine as long as they work with us."

Next, I shared my work with the senior operations leaders from our other international partners. All were receptive and appreciated the opportunity to voice their input.

After I had talked to the main parties, I took my analysis to Holloway for his information. He listened quietly before reiterating that this was my approach, not his. He reminded me that he would not fight the Russians—any problems that came up were mine to work out. When I told him about the agreement I had made with Blagov, he did not comment, and when I pointed out the hole where Jay was intentionally ignoring the Russian element, he shrugged.

The new work started to pay dividends almost immediately as I used it to resolve issues and validate our agreements. Soon, the Russians began to show some curiosity about what had been planned and why—especially what was required of each party. I suspected they were concerned about my fairness in allocating work and resources, so I emphasized that the MMT would publicly report everyone's contributions and obligations. The best antidote to mistrust is transparency.

As I made progress with the Russians, Tommy was struggling to find his footing with all the international partners. He was forceful—sometimes even harsh—when addressing issues with his American colleagues, but he left the internationals alone, avoiding potentially difficult topics. Based on our time together in Shuttle-Mir, I knew he was not comfortable working with people whose cultures and values were so different from his.

This boded poorly for the future of a program built on international cooperation. I urged him to create a senior position solely dedicated to building a strong bond with our international counterparts. We had an international office for handling

Management Chaos

Early in my time in Phase 1, I asked Holloway about his management structure. In reply, he took a piece of paper and wrote his name at the top, circling it once. Then, he began to draw swirling lines below this circle, describing the work as people "churning" until it was done. At the time, I thought this reflected the lack of structure in his office, but he later reiterated it while he was running the Space Shuttle program, a highly structured organization. Only then did I realize that it represented his literal view of how he saw the organization working.

protocols and budgets, but they were largely there to implement agreements made by the technical teams. I proposed that he find someone senior, with an operational focus and great people skills, to represent him as a senior liaison. I emphasized that this role should facilitate policy and technical decisions made by the experts, not add noise to an already crowded conversation.

As was his wont, Tommy heard part of what I said, then took action without sharing his plan. One day, without warning, he named Bob Cabana as his manager for international operations—a new position without clear roles and responsibilities that appeared to conflict with my job.

Bob, an experienced shuttle astronaut and commander of the mission that had delivered our *Unity* node to the *Zarya*, was well-suited to be a liaison. Bright, affable, and energetic, he was well-liked by everyone. But the uncertainty about his role instantly caused confusion that undermined the organization and trust I had worked so hard to create. Within hours, I received a dozen queries about Bob's authority and where he fit into our work. Bob was surprised and dismayed by my reaction to his role, as he genuinely wanted to make the program stronger. Despite his good intentions, the confusion sown by Tommy's vague charter continued to grow until Bob moved to the deputy of operations in Russia position—a job with well-defined responsibilities.

Much of my energy was focused on the Mission Management Team. The MMT process I inherited worked well with only a few modifications for its extensive international participation. The NASA team was well-trained to follow the direction of the MMT chair, and I was growing into the role. The Russians had participated in enough missions to understand how the MMT worked and generally went along with my decisions. But it stung their pride and sense of independent professionalism to have to take direction from another organization. They wanted NASA to accept their judgment, not second-guess them. The sting of Jay Greene's insults lingered for years.

My constant focus was to continue improving the working relationship with our most important MMT partners. My role in Phase 1 and my willingness to stick by Russia through that difficult time gave me credibility with them, but I wondered whether they would follow my decisions when the going got tough. If they did not, the only formal recourse was through the Space Station Control Board, which could not act quickly enough to deal with an emergency. I knew I had to keep the MMT's authority intact against the day I would need it in a crisis.

Not everybody was happy with my management style, which had evolved based on the history I had experienced. The problems on Mir had been made much worse by the lack of coordination across the team. As a result, I kept a close watch to ensure everyone understood and followed the plan. That led some people to feel as though they were being kept on too short of a leash. Fortunately, most people found our work satisfying. They knew they were part of a great team accomplishing challenging tasks and making history on a daily basis.

Some in the Space Shuttle program office, however, viewed the rising prominence of ISS as a threat to their status. They refused to accept ISS as an equal partner, doing everything they could to lord over us. Others acted as if they knew better than we did how to manage our hardware and plan our operations. They treated our launch packages as a payload subservient to the Space Shuttle program's routines while ignoring the challenges we faced in maintaining our own vehicle.

This was the same way they had treated us in Phase 1, when they had hurt some of our research and caused us a lot of extra work. Now, this treatment could put the future of the entire ISS at risk. NASA had committed to supporting ISS with all its resources, and everyone, including the Space Shuttle program team, had to do everything they could to do so.

I was particularly disappointed that this poor support continued when Bill Gerstenmaier became the Space Shuttle program's counterpart to my role in ISS. His experience in Phase 1 should have given him the insight and motivation to make the Space Shuttle program our partner. Instead, he let his flight integration managers continue to behave like petty bureaucrats. Some even told me that he was behind their passive aggression, though I did not believe it.

The Russians contributed plenty of pain as well, but they at least had an excuse. Their technical culture was different from ours, which resulted in some issues that simply would not go away. Most of these ended up with my MMT sooner or later, and I got lots of practice dealing with them.

24. MOD AND ISS

The Mission Operations Directorate and its people had changed my life by bringing me to Houston and baptizing me in mission operations. My feelings toward MOD went far beyond gratitude. I loved the people and the organization and would have been content to spend the rest of my career there. When others had decided that my skills as a system manager were more valuable to the station program, it had pained me to leave the organization. When those duties led me to work with them once again, I was delighted to renew my relationship with this outstanding team.

The leader of MOD, Randy Stone, had spent his entire career in MOD and was the head of the Flight Director's Office when I arrived at JSC. Despite his senior status, Randy warmly welcomed me into the MOD family and generously shared his time and experience with the newest guy in the directorate. I had been delighted when our relationship had moved to a new level.

By the time I took on the ISS operations role, Randy was in charge of the mission operations organization that supported both ISS and the Space Shuttle program. He was supremely dedicated, working tirelessly to make both programs successful, including by fostering good relationships with his counterparts in the international organizations. He had seen the challenges we'd faced working with the Russians on Mir and could foresee what lay ahead for ISS. I knew we would work well together.

But before we could move forward, we had to correct a problem from the past. One day, as we were ending a meeting, Randy asked for a moment alone. After closing the door to my office, he stiffened.

"I can understand if someone has a problem with my people. These things happen. But what I cannot tolerate is someone who criticizes them without talking to me first!" His eyes flashed with an anger I had not seen before.

"I don't know what you are talking about," I stammered.

"I have heard you with my own ears! You have said that you have no confidence in my flight directors for international missions, and that cannot be. If we are going to work together, you have got to come to me so that we can address any issues you have. You may be right, or you may be wrong, but my flight directors are the people who are going to make this work! So we need to come to an understanding of what needs to be done!"

My face flushed with embarrassment. I realized that he was right. The frustration I felt with the flight directors—who had resisted bringing the Russians into the program, criticized their ways of doing business, and made life tough for Frank and me—had shown more than it should have. More importantly, I assumed that he had known and had acquiesced in their behavior.

"You are right, Randy, and I am sorry. It will not happen again."

And it did not. I made a regular practice of dropping in to see him whenever I was on the eighth floor of Building 1, and he did the same for me when he was near my office. We spent many hours talking about what needed to be done to better support ISS. We reviewed the problems that had occurred during the Mir program. He was surprised but did not argue, accepted my input, and incorporated it into his planning for the future. He admitted that they had not always given us their best people and apologized for the problems that had created. In turn, I acknowledged that I should have come to him immediately and vowed to do better in the future.

From the beginning of our ISS work with the Russians, Randy sought ways to improve MOD's support for the new way of operating. He also understood that different skills and different people were required and that change would be inevitable. He spent many hours studying the problem and shared his insights with me. He was my most critical partner in developing an approach that would work.

His Russian counterpart was a former cosmonaut and flight director named Vladimir Solovyov. Solovyov was brilliant, outspoken, fiercely proud of his team, confident in his hardware, and mistrustful of everything American. He thought we were arrogant, spoiled, extravagant in our planning, and egregiously wasteful of the enormous resources at our disposal. Most of all, he had no intention of being subservient to spoiled Americans and even hinted darkly that someday, when their economy recovered, the Russians might just undock their segment and fly away on their own. Such an outcome would be a disaster for NASA, leaving the US segment without propulsion and other critical capabilities.

Randy and I talked at length about working with Solovyov and the other partners. He understood that I could not automatically take MOD's side on issues. He defended my efforts to listen to everyone's views, even when it incurred the wrath of his flight directors. He saw the scope of my challenges and did his best to support me; we were in this together. Most importantly, Randy accepted the idea that the Russians did some things better than we did. What mattered most was finding the best way, regardless of where it came from.

Randy also understood the mess Holloway had created and the toll it was taking on me.

"Your job would drive me crazy," he said, leaning back in his chair. "You are responsible for this incredible, high-risk enterprise with virtually zero authority over your biggest partners! Worse, your own boss is not fully supporting you. You are either very brave or very foolish!"

"Maybe both," I laughed. "At least I have one thing going for me. Even though people whine and play politics, both sides know that ultimately, there has to be a boss, and the org chart says that's me. If I am really careful and really lucky, that just might be enough."

"Let's hope so!" he said heartily. His warm smile showed his support.

"My biggest worry," I continued, "is that someone will force me to go to Holloway for support he will not give. That would be crippling."

Randy nodded knowingly. "Which is why you want me to get closer to Solovyov—to keep issues from going critical."

"Exactly. And it makes sense anyway."

"Of course it does. I will do my very best."

Eventually, as issues came up, we got around to talking about my problems with some of the flight directors. Both Randy and I had complete confidence in their technical knowledge, but we shared concerns about their non-technical judgment and how they interacted with others. He was horrified to learn that some had undermined our program by scaring away potential ops leads during Phase 1.

"You know, something like being an ops lead would have really fired me up when I was in my mid-career!" Randy said with a hint of embarrassment. "How could my guys be so narrow-minded? And how could I have missed it?"

We were both admitting our errors and committing to do better. Randy also did many little things to quietly show he respected the enormous challenge I faced and supported my leadership. One small step paid handsome returns over time.

Every organization has a hierarchy of minor details that indicate the flow of power through the group. Even though these details may seem trivial, they capture the attention of those who are ambitious. One such detail at JSC was the allocation of reserved parking spaces, particularly at major venues like the control center.

Shortly after assuming the leadership of ISS operations, I walked out of the control center and was startled to find my name on the very first parking spot next to the door. Randy's slot was next to mine, followed by JSC Director George Abbey, ISS Manager Tommy Holloway, Shuttle Manager Ron Dittemore, and other senior managers. In one fell swoop, Randy told the entire operations team that I was the man they needed to support and that he was there to make sure it happened. Nobody ever said a word about it, but I could see the effect immediately. People who had previously ignored me suddenly made a point of coming to see me about issues. Others made a point of informing me before making significant decisions. Both were welcome behaviors.

Randy and I grew closer as we worked through the constant string of challenges that came with operating our little station. One close call illustrates the growing pains that might have killed the baby.

One Friday, the air force unit that tracks space debris informed mission control of a possible collision with an old missile body. After refining the trajectory data overnight, it became clear that the debris would pass close enough to the station that we should maneuver to reduce the risk of a collision. First thing Saturday morning, MOD began its final planning. The duty flight director called to let me know what was going on. Having watched many similar events in the Space Shuttle program, I was not concerned, though I was curious about how well they were working with the Russians. The only rocket engines on the station were Russian, so we needed them to command the maneuvers, and since the space station was asleep at the time, it was not a trivial task.

What followed was a perfect demonstration of the cultural and technical gaps that could cripple us. NASA had had the air force constantly tracking orbital debris during every shuttle flight, and MOD had developed techniques to easily execute an avoidance maneuver with only a few minutes notice. Now, the air force was performing the same service for the new space station, and we had rules that dictated when we would maneuver the station to avoid the risk of collision, but we did not have a mature procedure to do so.

Because the Russians had little ability to track orbital debris, they never maneuvered to avoid it, simply accepting the small risk of a collision. As a result, their process for commanding maneuvers was slow, manual, and dependent on testing in a simulator before use. Knowing this from our time with our crews on Mir, I repeatedly told the flight directors to give them as much advance notice as possible.

My direction fell on deaf ears. Most of the flight directors had ignored Phase 1 and had only recently become involved in the space station. With a missile body bearing down on our precious station, they waited too long to ask the Russians for the maneuver, leaving them too little time to follow their normal procedure. The maneuver required a complex series of commands—to wake the station up from its sleep mode, move it to the correct attitude, then command the engines to give a specific velocity to the station. The Russians did their best to build the commands quickly, but they did not have time to run them through the simulator. As a result of being rushed, a minor error in the command sequence caused the attitude control system to shut down at the most critical moment.

Having been notified of the activity, Randy was in the control center to watch his teamwork. Even though his folks were confident that everything was going okay, he started to get uncomfortable and requested that I join them. I got there about thirty minutes before the maneuver and watched the situation develop.

The commands had been sent shortly before I arrived, and the flight control teams in Moscow and Houston were watching them execute their many automated steps. To their horror, the command suddenly aborted, and the attitude control system shut down just before the station went out of range of the ground station. This left us powerless to do anything about it or even track its fate. The next ninety minutes gave Randy and his flight directors plenty of time to reflect on how we had gotten to this awkward moment. I sat back and watched as they wrestled with what had happened and why. Fortunately, the station survived its close encounter, and the Russians regained control. The only real harm was to NASA's pride.

But there was an upside. As we were leaving the control center, Randy pulled me aside. "That was shameful. Mark my words—it will not happen again!" The pain and embarrassment in his eyes gave me confidence that it wouldn't. He later admitted that the experience had convinced him that the problems I had shared about Phase 1 were real.

The experience marked an important turning point for Randy. Even though he had listened to my concerns and done his best to prepare his team, seeing this failure put him in a much better position to guide MOD through the coming years. He would often reflect back on the moment as a painful lesson about allowing ourselves to become complacent. No lasting harm was done, and since the station had survived, we thought the embarrassment was private.

But somehow, the word got out. On Monday, ABC News called and wanted an interview to explain what had happened. Since it had been MOD's fault, Randy volunteered to take the heat for me, but Rob Navias wanted me to do it to show that the program was on top of the problem. Randy accompanied me to help with the heat if it got ugly.

As before, the interview would be with Ned Potter by phone using a camera crew from their local affiliate. It was a long interview during which I tried several different approaches to explain this complex subject to a lay audience. Ned thanked me for my help and assured me he would follow this one through until the story aired.

That evening, I once again watched the news as I worked in my office, but this time, I was startled to find it as the lead story. Instead of my explanation of the events, they used a clip of me saying, "Spaceflight is hard, and everyone makes mistakes. The key is to learn from those mistakes and do better next time. We are committed to doing just that."

At first, I was disappointed because I thought the technical explanation would have been helpful. But over the next few days, I got a lot of positive feedback that the quote had sent the perfect message. Spaceflight *was* hard, and mistakes *would* occur, even with NASA. My humility and commitment were seen as setting exactly the right tone for our new partnership.

That humility was heartfelt. The seemingly endless stream of assembly flights ahead was filled with risk, and we needed to move through them as quickly and as we safely could. To manage the risks, I pressed MOD to develop additional contingency plans. Some flight directors did not share my sense of urgency and attributed my push for steady progress to arrogance as if I viewed my work as more important than other things going on at the center.

They were dead wrong. Across my aviation career, both military and civilian, I had lost many friends to accidents. As a result, safety was always my number one priority, which led to some of our most contentious debates. There are three emergencies on a human spacecraft that require an immediate response—fire, decompression, and toxic spill. All require the crew to act immediately to protect themselves with little support from the ground. Mir had experienced all three during the time we'd had astronauts on board. While the Russians felt they had adequate procedures, most NASA people, including myself, were not happy with the way those events had unfolded. My intention was to be much better prepared on ISS.

Efforts to negotiate these procedures had been going on for years with only modest progress. With time running out before we would send up the crew, I made an extended trip to Russia to help force a resolution. As the one who would be held responsible, I wanted to be fully engaged. I also had experience with similar emergencies in military aircraft and felt I had useful insight to contribute. The negotiations were led by both NASA and Russian experts, and this time, they came up with answers that everyone could accept.

During this trip, I also visited Baikonur to see the launch facilities and examine the *Zvezda* service module prior to its launch. In my request for access, I explained that I knew exactly what I could ask the shuttle to do in an emergency and wanted to know the same for the Russian system.

The trip was set up by Dave Lengyel, one of our senior NASA people in Russia as well as a friend and former marine aviator. When I first proposed the visit, he said, "That is a great idea. It can be pretty rough at times, and you may regret it at first, but in the end, you will be very glad that you went."

Those words echoed in my head as we boarded an aging Tupolev Tu-154 airliner for the trip to remote central Asia. Wedged in my tiny seat, with my knees embedded in the back of the passenger in front of me, Dave broke the news that our return had not been scheduled and that arrangements might take a week or more. He also chose this moment to describe our crude living quarters where the workers stayed in Baikonur. While I had agreed with his recommendation to stay there rather than in a Western hotel, he had waited to share the details, afraid that I might back out. He was wise to do so, as my enthusiasm for the trip was in precipitous decline.

> ### Experience Matters
>
> To better understand the Russians' plans, I spent time with a senior systems instructor inside their training module. At first, he lectured me on their plans to respond to the emergencies with no interest in my views. This abruptly changed when I told him of my own experience with several aircraft fires, including my Swift accident. Hearing the voice of experience, he began to listen to what I had to say. The ultimate agreement was a compromise, but I was confident it was a good one.

The Baikonur Cosmodrome sits about ten miles outside the city of Tyuratam, Kazakhstan. An enormous reservation filled with desert scrub and aging facilities, it offers silent testimony to the Soviet Union's space program's once-grand ambitions. We stayed in a decrepit boarding house across the street from the neat dachas that had housed Gagarin and Korolev before the historic first manned launch. We were surrounded by the carcasses of large dormitories that had once housed thousands of workers. Now, their roofs had caved in, windows had shattered, and trees grew in former living spaces. The hardier industrial buildings appeared sound, but many had long ago fallen out of use.

What had been a busy place at the peak of the Soviet space era felt like a sad ghost town. It made me appreciate the accomplishments and dedication of those who had worked here.

Because of my senior role in ISS, I was granted permission to visit everything at the site, and I took full advantage of it. Some of the facilities were state of the art, while others were barely functional. All exhibited the brutal practicality that characterized Russian designs.

My tour of the enormous processing facility where both the Soviet Union's *Buran* space shuttle and moon rockets had been assembled offered proof of the remarkable capability they had developed. When we rounded a corner and saw the *Buran*, I was awestruck. Sitting on its Energia booster, with traces of reentry heat on its skin, it looked ready to launch again. I was heartbroken when, a few years later, it was destroyed after the roof collapsed under a heavy snowfall.

As we were leaving the facility, our guide spoke with great pride about what they had accomplished and hoped for the future. He was a serious man about my age, and I felt a professional kinship with him. Shaking his hand, I spoke to him with genuine warmth and respect.

"You have a magnificent facility here with much to be proud of. I hope that the partnership between our countries continues and grows and that we can someday work together on a rocket to take our people to Mars!"

He smiled broadly and said he would welcome me back.

When my chance came to see the ISS service module, *Zvezda*, its Russian manager warned me that this was a very busy time and that he was being very nice to let me in at all. He insisted that I could only have ten minutes and that the interpreter could not accompany me. Since he spoke no English, I knew we would not waste time on small talk. Even ten minutes would be enough for me to examine the workmanship and become oriented to its systems. I have always found that seeing the actual flight hardware gives me better insight into the challenges to come.

Once we were inside, I recognized the carbon dioxide removal system mounted on the starboard wall of the module. Wanting to show my guide that I was familiar with their work, I pointed to it and said, "*Vozdukh.*"

The manager lit up like a Christmas tree, grinning broadly and yelling, "*Da, da, vozdukh!*" He suddenly grabbed my arm and pulled me toward another panel, beginning a nonstop monologue in Russian.

We spent the next hour crawling all through the module as he excitedly talked about every face and instrument. I understood only a tenth of what he said, but I studied everything I could while nodding and saying, "*Da, da!*" He even pulled panels off and urged me to crawl down into spaces where I could barely fit. Since the design was nearly identical to the Mir's base block, I made special note of the areas where condensate had been a problem and the work it would take to clean it up while hoping ardently that would not be needed on ISS. I had great fun, and when we finally came out, I shook his hand warmly and thanked him sincerely for showing me his fine module. He beamed with pride.

Dave had been right. There were times when I hated my experience at Baikonur, but I still treasure it as one of the greatest weeks of my life.

25. NEXT STEPS

Preparations for the high-risk period of the assembly sequence were coming together well during the spring of 2000. This was known as ISS Phase 2—the critical time when performance margins, consumable supplies, and system redundancy were all at their minimums. This period had been my primary concern when I'd worked maintenance for the Freedom program. We had made important improvements over Freedom, and I was confident we had a good design, but it was still a very imposing phase. One of my biggest concerns was complacency, so I made sure the team stayed focused on the work ahead.

The urgent phase of assembly would begin with the launch and docking of the Russian element known as *Zvezda*, Russian for "star." This was the same module I had inspected at Baikonur that carried all the basic services for living in space to support the early crews. Because of its large size, the *Zvezda* could only be launched on the Russian Proton rocket. *Zvezda* would play a critical role throughout the program, and we were pleased to note that the Proton had matured into one of the world's most reliable rockets. Then, after a long string of successful launches, two Protons were lost within a year before *Zvezda*'s launch—both due to explosions during the second stage.

After the first failure, the Russian investigation concluded that the engine had been contaminated with dirt. They improved their quality control and gave the rocket a clean bill of health. Two successful launches followed, and everyone believed the problem had been fixed. But when an identical second failure occurred, our concern went through the roof.

If we lost the *Zvezda*, station assembly would halt until we could build a replacement. That would take at least three years, during which time political support and other issues could very well kill the program.

With the entire ISS at stake, NASA insisted on being a part of the investigation of the second launch failure. A joint review commission, cochaired by former astronaut Tom Stafford, thoroughly examined the Proton's design and operation. The commission included my friend and former boss, Michael Greenfield, whose expertise as a metallurgist was invaluable in restoring our confidence in the rocket. He was pleased with the work the Russians had done and marveled at how robust the engines were. They traced the problem to manufacturing issues caused by financial stresses at the factory and fixed it. Our participation restored our confidence that the engines were ready for this critical launch.

After years of hard work, we were finally on the eve of the most urgent phase of the program—what I liked to call the "Big Push." During this time, the stakes would keep rising as we added hardware and crew. We would be in a race to launch, assemble, and activate enough systems for the station to take care of itself before an inevitable problem arose.

The final weeks before the *Zvezda* launch reminded me of the start of a roller coaster ride—when the cars are pulled to the top of the incline just before the plunge. Our wild ride was about to begin. Win or lose, whatever came next would change our lives forever. A successful launch would begin the torrent of activity that could not be stopped. Once we attached the service module, we could no longer put the station to sleep. We would have to live up to our promise to support it continuously for at least fifteen years—hopefully much longer.

Despite the difficulties of international cooperation, the decision to add the Russians had given us two sets of vehicles and flight control teams to share the workload. Our Phase 1 apprenticeship on Mir had taught us a lot about supporting long-duration missions and the Russian way of doing business. The months spent flying just *Unity* and *Zarya* had given us time to build our partnership and refine our processes before the truly risky work began. Now, Phase 2 would push us forward into the delicate process of assembling the station—bringing the full measure of risk to our rapidly maturing team. I dearly hoped that we were ready.

As the *Zvezda* launch approached, we conducted joint reviews even though the work was entirely in Russian hands. Modules similar to *Zvezda* and *Zarya* had successfully docked many times, so we did our best to patiently watch the Russians do their thing.

Being passive was not easy for any of us. After the module's successful launch, the Russians took two weeks to check it out and make the approach for docking. As the time crawled by, my anxiety about the docking grew. Randy Stone could see the tension and graciously offered me a seat at the console immediately behind the flight director and alongside Milt Heflin, one of the most senior managers in MOD. Although the role of the Houston mission controllers remained passive, it helped to be in the loop.

The docking went well with only one small issue noted. A video of the *Zvezda* taken from the *Zarya* showed that one of the panels on a solar array had not been fully deployed. The Russians indicated that this was not a problem and could be fixed on a future EVA. Knowing this panel would have to be moved at some point, I tried to ease the Russians into our style of planning by getting them to schedule the operation. However, my routine inquiries about the solar array didn't generate interest, so later that day, I decided to call Victor Blagov.

"Victor, congratulations on the successful docking! It was beautifully done."

"Thank you, Jim," he said through the interpreter. "It is good hardware, and it worked well."

"I noticed that one solar array panel didn't deploy. If you want to have that taken care of during an upcoming EVA, I ask that you make a request to our planners. We like to know what we are going to do before we send the crew outside."

"It is not important. We have plenty of power from the array. We will fix the panel when it is convenient."

"As you like, Victor, but please don't wait until the last minute. Let's incorporate it into our plans."

Despite my request, Victor never mentioned the solar array during any of our planning sessions or MMT meetings.

A few weeks later, the Russians launched a Progress resupply vehicle. The craft successfully docked with the aft end of *Zvezda*, bringing supplies that a shuttle crew would unload a few weeks later. The shuttle would also bring a Spacehab double module crammed with the logistics needed for the first ISS crew to activate the station. The shuttle crew would unpack both the Spacehab and the Progress, remove packing materials and other trash, and perform an EVA to prepare the station for its first permanent crew.

Shortly after that shuttle had launched, the Russians brought an old issue back to the MMT. The shuttle had a design feature that created a small risk of a control jet firing without being commanded. In free flight, this was a minor problem, but when the shuttle was docked to another vehicle, it could break the airlock, which connected the shuttle with the station. If the jet continued firing for more than a few seconds, the broken airlock would vent all the air out of both vehicles, killing everyone on board. On missions to Mir, we'd kept an astronaut at the controls to rapidly shut the jets down if they fired. Because it was our hardware, we were comfortable that the risk was minuscule.

But the Russians were not as comfortable, and they wanted to review the issue one more time. Normally, such issues would be worked through the engineering teams, but with Jay Greene's hostility toward the Russians, they preferred to bring the issue to me. It soon became the topic of a formal MMT decision.

Although I was tired of going over this familiar issue, I was proud that the MMT was respected enough to review the issue in front of all parties. Talking about a familiar issue over and over again was as frustrating for me as for anyone, but I was committed to taking as much time as needed to make the Russians comfortable. Even though the risk of failure was exceedingly small, the consequences could be catastrophic, and managing that risk deserved our full attention. After more than an hour of discussion, we got a Russian agreement to continue.

When we finally completed the meeting, Jim Halsell, an experienced astronaut and the shuttle program's deputy for KSC operations, came over.

"Jim, you feeling okay?" he asked.

"I'm fine—just tired," I said. "Why do you ask?"

Jim looked at me thoughtfully before saying, "You need to take care of yourself! I am here to tell you that *nobody* wants your job!" He laughed when he said it, but we both knew he was telling the truth. There were youngsters who lusted after the "glory" of my job, but no one with the experience to actually do it did. Many thought I had been set up to fail. At least that meant I had job security!

The shuttle flight included an EVA to connect some electrical cables between the Russian modules and perform other assembly tasks. I watched the first two hours of the EVA, which was occurring in the middle of the night in Houston, from the control center. Since everything was going smoothly and the next day would be busy, I decided to go home and watch the rest of the activities on NASA's TV channel in hopes of catching a few hours of sleep.

EVAs are extremely intense for those involved, but for others, they can be about as exciting as watching paint dry. Despite my best efforts, I fell asleep in front of the TV.

I was awakened by my cellphone. The voice of the flight director got my full attention.

"Vladimir Solovyov was calling for you. He wants to send the EVA crew over to extend the stuck solar array panel. We are slightly behind the timeline, but our guys don't think it would take too long. We could do it if you approve, but we don't want to."

"What about Harbaugh? What does he say?" Greg Harbaugh was an experienced EVA astronaut who now ran the EVA project office for JSC.

"I'm not sure. I haven't seen him."

"How long do we have to decide?"

"The crew is cleaning up. They will be back at the airlock in thirty minutes or so."

"Okay, tell Solovyov I will call an emergency MMT telecon to discuss it. I'll get you an answer in fifteen minutes. Does that work for you?"

"That works."

I called my team and told them to activate the emergency MMT process. This was an outgrowth of what we had done during the bio-reactor problem on Mir. It allowed us to get critical people together on a "meet me" phone line within minutes, no matter where they were. I had been looking for a chance to exercise the process during an assembly flight, and this was it.

Five minutes later, we had the whole team assembled on the phone. Most of them were in our MMT conference room, but we also had representatives in Moscow and elsewhere who were tied in by phone, as I was.

Using the same process as a normal MMT meeting, I went around the virtual table, asking each party to address the request. The idea to add the task had occurred to the Russians while watching the crew working close to the solar array. The Russians had not made any preparations for the task because they had not thought any were required. Harbaugh, our expert, felt it was probably a simple task, but he did not know. He had no detailed pictures and could not vouch for what might be required.

When I polled the members of the MMT, everyone except the Russians was opposed or neutral to adding the task in the middle of the EVA. My decision was easy: no-go.

I also took the time to explain what we on the American side thought was necessary for a safe EVA in the hopes that Solovyov and Blagov would provide us with more lead time for planning. But Solovyov had left the MMT early when he'd seen the tide turning against them, and Blagov acted as if I were making a mountain out of a molehill.

We had invested countless hours into planning the assembly process, but our progress was often hindered by those who were not so forward-looking. Even people who should have known better often interfered with our work without trying to understand what we were doing and why. This was not the first time, and it would not be the last.

26. FINAL PREPARATIONS

As the summer of 2000 wound down, the atmosphere was electric. For the first time in my life, I was a part of a team that was about to start something truly audacious—during which my leadership would play an integral role in our success or failure. One part of me was terrified that I might let the team down, but another part was quietly confident. Many nights, I lay awake wondering which part was more accurately assessing the future and how I should manage my confidence and anxiety to best prepare for what lay ahead.

There was no question about the technical challenge of our work. My bookshelf was jammed with 35,000 technical requirements and countless plans for the assembly operations. My days were spent tending our tiny station in the mornings and reviewing plans through the afternoons and evenings. The work was hard, but our progress was steady—we were really going to do this!

It was time to turn our station from a sleeping baby into a rapidly growing adolescent.

My biggest challenge was human behavior. The Russians had been flying their space stations for nearly thirty years, the shuttle had been flying for almost twenty, and some parts of the American space station had been under development for nearly fifteen—giving both space agencies plenty of time to get stuck in their ways. Nobody wanted to change, yet each program would be forced to adapt as they worked together intimately.

Because I had moved through three different programs—the Space Shuttle program, Space Station Freedom, and ISS—in less than a decade, accepting this change was easy for me. My work in Phase 1 had immersed me in Russian culture and operations, crossing an enormous cultural divide in a breathtakingly short time. It almost felt as if I were being guided by a destiny I could not foresee. Now, it was my responsibility to blend those experiences into the best path forward for our new program.

One of many concerns was the knowledge that Ryumin and his colleagues viewed ISS not as a long-term commitment but as something to tolerate until their space program got back on its feet. This was both an advantage, in that it got them to participate, and a liability, given the half-hearted effort some of them applied.

This framed my most urgent problem. We needed the teams to work together closely, sharing information and helping each other without reservation or delay. As we prepared to launch the first permanent crew to the ISS, I had no choice but to trust that the agreements were sound.

That confidence was deeply shaken when Tommy Holloway came by my office one Monday afternoon in late September. He told me that he was going to talk to JSC Director George Abbey about an issue with the Russian leadership and that I should be ready to head to Moscow very soon.

With a busy week ahead, I asked, "How soon?"

"Tomorrow," he answered reluctantly.

"What is this about?" I asked.

"Seems there are Russians who don't like your leadership style."

"No surprise there," I answered. "They're the same ones who don't like working with NASA, but that ship has sailed. The agreements I negotiated last year are based on what they themselves told me they need to do the job. If they're complaining about that, I need to know. I'll come with you," I said, but he shook his head.

"No, it's between George and me. You can wait in the outer office if you want, but I want to discuss it alone."

It was hard to imagine a reason why he would tell me about the discussion only to block me from attending—unless it was to taunt me. It took everything I had to control my anger. The boss who had thrown the whole burden of operations on me now seemed intent on undermining my confidence and authority. I was mighty tired of his leadership style.

As I sat in the reception area outside George's office, hoping that I would be allowed in, Mike Foale and others walked right in and joined the conversation. I thought about just barging in myself, but my military background would not allow me to violate a direct order. Though I couldn't bring myself to do that, I was sorely tempted to walk out and go on leave for a week. It took all the emotional strength I had to stay put and wait for the outcome.

When Tommy eventually came out, he said, "You need to go to Russia tomorrow."

"To do exactly what?" I asked.

"Fix whatever is wrong. Get the Russians back in the box."

"What are you reacting to? What is the problem I'm supposed to fix?"

Tommy simply shrugged.

Never in my life had I felt such anger and disgust with a boss's behavior. If not for my sense of responsibility to the program, I would have resigned on the spot. But the memory of the Cold War—and the precious opportunity the ISS offered to repair some of its damage—would not let me quit.

The vision of a trusting partnership gave me the strength, despite my anger, to focus on the work ahead. I knew that if Russian complaints had made their way to George, there was probably something that needed my attention. I just wished the Russians had talked to me first and that my leaders would be part of the solution rather than adding to my problems.

The flight to Moscow was long and lonely, filled with doubts and anxiety. Arriving too late Tuesday to meet with anyone that afternoon, I went directly to the hotel. After a restless night, I headed to the Russian mission control center.

My first stop was the NASA area. I asked our team if they knew what problem had led to my trip, but they didn't have much information to share. They knew that Solovyov was angry about the EVA task I had refused to add at the last moment, but they believed that was just the tip of the iceberg. Most thought the Russians were simply upset about no longer being in charge.

Then, as I was leaving, one of them added this:

"Solovyov was surprised to hear that you had gone home during the EVA. He couldn't believe that you weren't in the control room."

Filing that away for the moment, I headed upstairs to meet with the head of Russian mission operations, Vladimir Solovyov. He met with me and the lead flight director, Victor Blagov, in his office. He did not waste time on pleasantries.

"The way NASA does business is not a good way to run operations," he said. "You take too long to make a decision, and you let too many people get involved. You know you cannot make dinner with too many cooks. You need one chef!"

His manner was direct but not antagonistic. Everyone knew that Vladimir did not like Americans, but he seemed to be making an effort to be respectful. I listened carefully.

"It is our job to make decisions during the mission," he continued. "As a flight director, you know that we have to be ready to make those difficult choices. You cannot give that responsibility to others."

I did not correct his belief that I was a flight director; I suspected that all the leaders in their operations organization had been one. Perhaps it would have helped if he had known of my role in the design of the vehicle or my experience as a fighter pilot, but I doubted it. I wished he understood my responsibility to integrate and lead this immense international team, but he clearly did not.

I sensed something deeper going on. Although he did not say it, I believed this conversation was Solovyov's reluctant but tacit acceptance of my authority to make the final decision. From his manner, I suspected he was trying to lean on me to defer more decisions to him.

"I know it's my responsibility, and I have made many difficult decisions in my life. But I have seen far too many mistakes made because people rushed into a decision without asking for all useful input. That is why the NASA way is to listen to everyone we can and try not to make mistakes by rushing. My decision to call the MMT was not the reason I did not add the task to the EVA. I knew exactly how much time we had to make the decision, and we could have added your task if the team had agreed, but they did not."

"You need to be able to react! You need to be ready to take action when the opportunity presents itself," he said, leaning forward. "Life does not always go according to our plans!"

I leaned forward, too, and looked him straight in the eye.

"I agree completely, Vladimir, and I am always ready to do so when circumstances change. But nothing changed during that EVA except your last-minute request to add something without warning. We all knew about the solar array for a month, and yet you said nothing and did nothing to prepare to address it. Through the MMT and other channels, including speaking directly to Victor, I asked repeatedly if you wanted to add it to the EVA, but nobody said a word. Even during the mission, I heard nothing until we were wrapping up. If you had suggested it even as late as the beginning of the EVA, I would have been glad to do it. Your hardware is a critical part of our station, and we want it to be healthy. But we must consider the risks as well as the benefits of every action we take. We need to think ahead!"

Solovyov was not impressed. Obviously, he did not see the concerns that we did, or he was willing to accept a higher risk than we were. He got up to leave.

"I hope NASA can keep up with the challenges we will face," he said.

After Solovyov left the room, Victor and I went to his office where we continued the conversation. We had rarely talked one-on-one, and I suspected that he had something unpleasant to say. After a few moments, he took a forceful tone.

"Jim, we do not have as many people as NASA does, and we cannot waste as much time as you do. You should have let us fix the array. If you did not want the NASA astronaut to go, we would have sent Malenchenko." He was the Russian Cosmonaut who'd been a part of the EVA crew. "He could have done it easily. We are more flexible than you are."

"Victor," I said carefully, "you know that I respect and value your experience and advice. I have watched and listened carefully all through our time on Mir. Now I need you to listen to me."

Victor seemed surprised by my intensity. He sat back as I continued.

"As I told Vladimir, the NASA team did not say that they did not want to do the task. They said they did not want to add a task in the middle of an EVA when it could and should have been discussed ahead of time. This was not an emergency. It was not even a surprise. You have known about this solar array for more than a month, and yet you did nothing about it until the very last minute. That is not the way we work in NASA, and it is not the way I am going to work. When you have an idea for something you want to add or change, bring it to me as soon as you can, and I promise we will work hard to accommodate it. Wait until the last moment, and you will be disappointed."

Blagov looked pained. "We can't always think of these things in advance," he said. "Sometimes we have to do what we see is right, even without all of the planning."

"Victor, I take this job *very* seriously. We are facing many risky operations in the years to come, and I am doing my best to look out for your needs as well as ours. We cannot be successful without you, which is why I asked you about that solar array as soon as we saw it a month ago. All you had to do was say you wanted the option to fix it, and we would have jumped on it. We plan ahead because we have learned that it is the way to prevent serious mistakes.

"We are busy too. We have more people, yes, but we also have far more to do. If you took the time to study the design of the station, you would see that. I helped design this machine, and I know exactly what it requires from everyone. It is my job to be responsible for the whole operations program, not just NASA's. It takes every waking moment of my life, including many that should be spent with my family. I spent the entire day before the EVA in technical meetings reviewing the coming assembly missions, and I did the same the day after the EVA. All that planning allowed me to trust the team to do their job even when I went home for a few hours of rest. I understand that is not your way, but I have seen the results of your way. The decisions that resulted in the Progress hitting Mir were very unprofessional."

Victor started to speak, but I quickly raised my hand, and he stopped.

"In the US, such a lack of planning would be considered criminal negligence. If something like that happens in this program, someone is going to jail, and I promise you, *it is not going to be me*!"

Victor looked at me for a long moment before speaking.

"Okay, Jim," was all he said.

When I got back to Houston, I told Tommy and George what had happened. Tommy was unimpressed, but George gave a sly smile. He saw the progress we had been counting on.

A few days later, I got a surprise visit from Bill Shepherd, who would soon command the first crew to occupy the ISS.

"Hey, Shep, how's the training going?"

"Thankfully, it's almost over," he said. "It's been a long run."

"We knew that going in, but I am hoping it will be a little easier for later crews. So, what brings you here today?"

With a resounding thud, he set a Makita cordless saber saw on my desk.

"I want to fly this," he said.

"And why is that?"

"You never know what you will need to do in space," he said. "I want to be able to make adjustments without having to wait years for people to redesign things. And I want to make some things out of excess materials that are on board."

This confirmed some rumors I had heard—that the crew of the last shuttle mission had not disposed of all the structural support panels and other items that had been removed from the *Zvezda* during their mission. They were all supposed to have been put in the Progress for destruction, but there were reports that some had not.

"Has it been through the safety review process?" I asked, even though I knew the answer.

"No, it's mine from my shop."

"Have you discussed the dangers of operating a power saw in zero gravity? Have you thought about how you would anchor the work and control the dust and other debris that would be created?"

"No. I can deal with all that stuff. I want it because I can do things a million times faster than waiting for the ground to develop a solution."

I was not impressed.

"I agreed to certify and fly the drill and other things you asked for because I can see a legitimate need for a limited set of quick repairs or modifications," I said. "I did that only because you promised me you would not drill anywhere that had not been approved. Now you want to add an uncertified power saw to the mix."

Shep could see where the conversation was going and started to get up.

I walked him to the door.

"You are asking for a lot at a time when we are buried in work and our partnership is struggling to get along, but I will take this up with the team. And you are doing what I just chewed the Russians out for doing—bringing up with a new requirement at the last moment even when you had been working on it for a long time."

"You could learn a lot from watching the Russians," he said.

"I know, and I have," I said, "including how to nearly destroy a space station with a foolish act. You would benefit from being more critical of such foolish actions. Their experience, including the Progress collision, is proof positive that rushing into things brings more risk than we can tolerate."

He shook my hand and turned to leave.

"We are doing everything we can to make your expedition as safe, pleasant, and successful as possible. You need to trust me on that, and I need to trust you to do the right thing."

"No worries," he said as he walked away.

I called Gary Johnson, the head of our safety team, and told him about Bill's request.

"That is a terrible idea! What is wrong with Bill—he knows better. And even if we agreed to do it, there is no way the Russians would allow it. Please don't agree to it!"

"That is my assessment. Thanks."

I called Bill and told him the answer was no.

"I don't want to see that saw on orbit!" I said emphatically.

"No worries" was Shep's reply once again.

Although that was the last time I would hear about the saw while I worked for ISS, it was not the end of the story. After failing to get permission to take the saw, Shep asked a colleague to have it smuggled aboard. The first few crews managed to keep it a secret, but years later, it showed up and caused a major flap. I was disappointed but not surprised by his behavior, as many astronauts and cosmonauts felt entitled to do whatever they wanted.

27. WELCOME ABOARD

As the summer of 2000 turned into autumn, the ISS roller coaster plunged ahead. The excitement was palpable, energizing everyone to execute our amazing plan. We were soon in a headlong rush toward our future.

In early October, the next shuttle flight—what we called flight 3A—launched to deliver two more critical components of the core station: the Z1 truss and a pressurized mating adapter. This would be the last shuttle flight before Bill Shepherd and his crew arrived.

The Z1 was a short segment of truss that housed several critical systems. The most important of these were the control moment gyros, or CMGs. The CMGs are heavy gyroscopes that can hold and change the station's attitude without using propellant. They would not be operational until the *Destiny* laboratory delivered the computer system, but this was where they fit in the assembly sequence.

One day, toward the end of 3A, while I half-listened to the comms link between Mission Control and the shuttle from my office, I noticed the flight director who was serving as the lead for the flight starting to act strangely. The flight director—I will call him Chuck—began talking directly to the shuttle crew, bypassing the Capcom, which was a breach of our procedures. His unusual behavior caused me to listen more carefully, and I soon heard him exceeding the limits of his authority by deviating from the prelaunch plan in ways that should have been addressed by the MMT.

I knew Chuck from my time working with him during my detail to the Flight Director's Office. He was an on-orbit flight director who had been trusted with some of the most technically demanding missions of the day. His intelligence and energy were well-known—as was his enormous ego. A big ego was often part of

> **Names for ISS Elements**
> The different elements of the ISS were named based on either their milestone status or their location on the finished station. Major modules got individual names to note their value, while more prosaic elements had a synoptic name based on their final location. Z1 was the first element installed on the top (zenith) of the *Unity* node.

the package, especially with such aggressive and confident people. The trick was to channel that energy in the right direction—not an easy task, but one that had mostly worked for the Space Shuttle program. The difference on this flight was that he was ultimately responsible to the ISS program, not the Space Shuttle program. He was intentionally testing his limits.

I called him up. Chuck was obviously expecting my call as if he had been baiting me. He quickly unleashed a flippant and arrogant rant, clearly intended to insult and challenge me. It did, prompting a reaction to argue, but I quickly saw that was exactly what he wanted. After a moment, I regained my composure and told him to "knock it off and get back to work," or there would be consequences. He mocked me and laughed.

I slammed down the phone and tried to calm down. After a few moments, I called Jeff Bantle, the head of the Flight Director's Office. His secretary could tell from the tone of my voice that this was important. Within seconds, Jeff was on the line.

"Jeff, we need to talk! I'll be there in three minutes!"

"Okay," he said quietly.

We were in adjacent buildings connected by a walkway. My anger must have shown on my face as people literally scattered when they saw me coming. Jeff was waiting when I reached his suite. He waved me inside and closed the door.

Like most MOD leaders, Jeff usually radiated great confidence. On this occasion, however, his bearing was measured and tentative, as if he was facing a new problem for which he had no ready answer. Apparently, he had also been listening to Chuck on his speaker and noticed the odd behavior, but its significance hadn't registered until my call. Once alerted, he'd called Chuck to see what was going on. That short call had been enough for him to understand why I was angry. When I relayed the details of our conversation, he was shocked.

"I don't know what's gotten into him," Jeff said, "and we'll have to get to the bottom of it after the flight. But I'm sure you understand that I need him to complete the mission. As lead, he has all the details."

"I know how it works, Jeff," I said, "but I've frankly lost confidence in his ability to make responsible decisions. He has already exceeded his authority, and that cannot happen again!"

"He shouldn't have done that," Jeff said solemnly, "and I'll make sure he doesn't do it again. I'll put a senior manager on the MOD console to keep him in line."

"How you do it is up to you, but this *must not* happen again." Jeff could see that I was genuinely concerned about the success of the flight, not just angry at Chuck's insults.

"I'll get with Randy immediately, and we'll make sure it doesn't."

"Thank you. I know you have a lot on your plate, but this goes to the core of MOD's authority and responsibility to the programs. I need to be able to count on your guys doing what they're supposed to do."

"That's my job," he said. "We'll make it work."

Once again, MOD had shown itself to be a quality, professional organization. Jeff was clearly embarrassed that one of his senior people had gone so far astray and shown such poor judgment. Although we never spoke of the incident again, that was the last flight for Chuck, who went on to manage some technical projects elsewhere in the agency. Having known this man for nearly a decade, I was used to his histrionics, but I had always thought I could count on him to be professional. His bizarre behavior on this occasion made me wonder if we were asking too much of our people. I would find out soon enough.

Not long afterward, about a week before the new crew was set to launch, I had another visit from Randy Stone, director of MOD.

"I have some bad news. Although this isn't a good time for it, I've had to make some changes to the team. Two of my folks haven't been getting along with their Russian counterparts, and I just can't have them hurting our relationship."

"Who are we talking about?"

"The flight data file and planner positions," he said. "The problem we had with the maintenance procedure on 2A.1 was not an isolated incident, and the planner is just not getting along with his Russian counterpart. With Shep and his crew launching next week, we need to fix these problems ASAP. I hate making one change so late, let alone two, but we have no choice."

This was distressing news, but at least Randy was fixing it.

The first problem involved the flight data file—written procedures kept on the spacecraft. Flight 2A.1 had been the first shuttle flight to our baby space station after we had docked the *Unity* node to *Zarya*. When the crew pulled out the maintenance manual stored in the *Zarya* module, they found it was all in Russian rather than the dual-language format we had agreed to. At the time, we thought they had somehow launched the wrong version, but Randy learned that the American had not cooperated with his Russian counterpart to produce an agreed-upon bilingual version. As a result, the Russian had sent up the version he'd had confidence in rather than the contested dual-language version.

I was aware of the wrong file making its way on board the station because it was ultimately returned to me. The second problem Randy mentioned, however, was news to me. The head of our planning team—responsible for the daily plan the crew would execute—was having trouble with his Russian counterpart.

"Is the planning problem on our side or theirs?" I asked.

"There's no way to know for sure," Randy said. "You know better than I do that the Russians can be difficult to work with, but we have got to find a way. The chief Russian planner is a woman who is a close friend of Vladimir Solovyov. She wants things done her way with little compromise. Our guy has to be able to work with her, and the current guy is struggling."

My first impulse was to stand up for my NASA colleagues, but if Randy said we needed a change, I believed him.

"Who do you have in mind to take over?"

"The flight data file problem is relatively easy to manage. Most procedures get built over weeks or months, so there's usually time to work things out. The real problem is planning because it goes on night and day with the team constantly reacting to events. We're bringing in one of our top Space Shuttle program people, Scott Curtis. He has done an outstanding job there, and I believe he's our best choice to get the process back on track. I've already talked to him, and he's willing."

"Thank goodness for that! If you have confidence in him, I do too. But I'd like to meet with him to share my experience with the Russian planning process on Mir. He needs to hit the ground running so we can be ready when the crew gets there."

Scott was one of many people who took a leap from a secure job he knew well into the unknown of ISS. The Space Shuttle program was entering one of its busiest periods, and nobody would have thought less of him if he had stayed put. But Scott loved NASA, and when he was asked to help, he stepped up. He was another example of someone who came when called.

Scott's job was particularly challenging because its products would be used every day for as long as the ISS flew. He and his Russian counterpart would jointly produce the plan for what the crew would do each day, including last-minute changes when required. It was a demanding job that could consume every hour of every day.

Over the next year, Scott and I would spend an enormous amount of time together as he worked his way through the issues. He understood that I needed to let the Russians have their way on certain strategic issues. He only asked me to intervene on rare occasions when he thought it was essential.

The first crew—Shep and two cosmonauts—launched to ISS from Baikonur on October 31, 2000. The docking of their Soyuz capsule two days later marked the beginning of a period of continuous crew presence with no planned end date. As of this writing, more than twenty-four years later, ISS remains healthy, active, and productive.

The crew, all experienced space fliers, looked happy and comfortable as they settled into their new home. The near-death experience of the 1993 redesign made this moment feel like a triumph. We allowed ourselves a few minutes of celebration before getting back to work.

The mismatch between the established organizational culture and the needs of ISS was not limited to Russian participation. Phase

---

Lessons from the Past

This experience added to my already-deep interest in lessons learned from other programs. A little over a decade before our problem, NASA had had a problem deploying the high-gain antenna on the *Galileo* probe. That problem had been caused by a combination of age and cost-cutting changes to its transportation plan that allowed lubricants to erode and prevent the antenna from releasing completely. The antenna had never deployed completely, though the engineers had found a way to recover 70 percent of the science data by other means.

1 had exposed disconnects with the Space Shuttle program culture. The Space Shuttle program's deeply rooted procedures and previously unchallenged authority over manned spaceflight clashed with ISS's need for flexibility and its own sphere of authority. The stress test of Phase 2 soon brought this into stark relief.

With full-scale operations underway, the ISS roller coaster was running at full speed. Despite all our preparation, reality soon exposed many weak areas. New issues arose every day, and the team and MMT did their best to address the most pressing ones expeditiously. My Freedom experience made me acutely sensitive to the need to get through this vulnerable time as fast as safety allowed. I pushed people to find workable solutions we could implement immediately rather than perfect ones that took time. Some resented the pace I set and the degree to which I insisted on being involved. Their reaction was disappointing. We all had to be on our best game to make it through this crucial time.

As soon as our first crew settled in, shuttle flight 4A delivered the first American power module with its two enormous solar arrays. This module, called P6 for its final position on the port end of the truss, would provide the electricity needed for the first American elements. In one of the many temporary configurations required by the assembly process, the module had to be mounted on top of the Z1 until its permanent location was ready to receive it.

Each array was made up of flat solar panels that folded like an accordion, allowing each 110-foot-tall solar array "wing" and its mast to be folded into containers only a few feet thick. As we watched the first solar array's deployment in mission control, we saw that instead of its sections unfolding evenly, as expected, the flat plates seemed to stick together, then break loose in violent spasms. The sticking panels caused a lot of tension to build in the mechanism, and when a panel broke free and unfolded, the energy released caused the panels to jerk violently.

It took us a few minutes to appreciate the possible consequences of what we were seeing and stop the deployment. Our first concern was that this might break the connections between the panels, so we immediately assembled the experts in the Mission Evaluation Room (MER).

Because the solar array deployment involved Space Shuttle program crew and equipment, the Space Shuttle program operations manager, Bill Gerstenmaier, was technically in charge of resolving the problem. This was true even though I owned the ISS hardware and systems expertise. Bill listened to our discussion and agreed with the theory that the extra time the solar arrays had spent folded up in storage had caused the panels to stick together. The delay had been caused by budget and technical setbacks that had left this set of arrays in storage for much longer than intended.

Our engineers recommended continuing the deployment slowly, allowing time for the sun to warm the panels and hopefully break the adhesion between them. This approach, combined with moving the panels in small intervals, worked well. We were able to extend the first array with no obvious damage, but the jerking movement had made the deployment cable jump off its pulley. Now fully deployed, the array couldn't be rigidized and remained floppy and vulnerable to damage.

We had no procedure to fix this problem. Although we were confident that we could find a solution, we needed some time to think it through. In the meantime, we had to decide what to do with the other array. The schedule called for extending it immediately after the first one, but the problems we had encountered made us reconsider.

We had every reason to believe that the second array would behave the same way as the first one. The modified deployment procedure seemed to have worked, but there were countless details we did not understand, including how much damage might have already been done and how to repair the damage to the tensioning mechanism. Moreover, the time we had spent on the first array had put us way behind schedule. This was putting pressure on the team to get it done before the end of the crew day. To make sure we did not screw up the second array, I wanted to wait until the next day to deploy it.

Bill did not. He wanted to use the new procedure and get the second array done. Although this was logical in some ways, it also put us at risk of behavior we did not fully understand. I argued to delay, but Bill dismissed my recommendation without giving his reasons and pushed on.

This was one of the few times in my NASA career that I was truly angry at a colleague. Bill was putting my hardware, program, and career at risk. And even though he had spent some time studying the hardware, he was acting over the objection of the individual most responsible for its future—me. This was extremely unusual for NASA and, in my view, very poor judgment.

The second array did deploy successfully. Warming the arrays worked and reduced the disturbances enough that the tension cables stayed on their pulleys. Despite the success, I knew that our success had depended on good luck rather than good decision-making—a very bad combination.

With both arrays deployed, we needed to repair the tension system on the first array. An EVA crewman was soon airborne in a NASA T-38 jet on his way to the manufacturer's plant. There, he examined the way the tensioning cables were installed and developed a plan for getting them back on their pulleys. The next day, the MMT reviewed the technique they had developed and gave its approval for the crew to perform the task during the next EVA. It worked beautifully.

The rest of the mission went well, and it looked like the year 2000 was on track for a successful conclusion. The crew was happy, the ship was working well, and the next element to launch, the US *Destiny* laboratory, was ready to go in January. There were lots of little issues being worked on, but the outlook for the station was bright.

A few days later, Tommy Holloway came to my office.

"You need to take some time off," he said. "You work harder than anyone I have ever known, and you need to get some rest."

Tommy had a habit of exaggerating to make a point, but I was pretty tired.

"I appreciate the thought, Tommy, but there are a lot of loose ends that need my attention."

"Nobody is indispensable," he said. "The team can handle it. I want you to take time off."

"My wife would love that, and I will if I can."

He looked at me hard. "Make it happen!"

Although unconvinced it was wise, I started planning a short vacation. My wife happily greeted the news and reported that her parents would love to have us visit them in Michigan. We began making plans to fly up in our Twin Comanche, but even as those plans were coming together, problems were building on the station.

Every morning began for me before six o'clock when I arrived at the Mission Evaluation Room. The MER was located in the basement of the Mission Control Center, where its engineers sat at rows of consoles amid stacks of manuals and reference documents, monitoring the station's various systems to assess their performance. The walls displayed running lists of things we did not understand and that needed attention—what we called items for investigation (IFIs) and in-flight anomalies (IFAs). We typically had several dozen issues being tracked in those early days, mostly describing what we thought were minor performance issues. All were treated seriously, and my biggest worry was that several small issues could combine to make a big one.

That worry began to grow as we noticed unexpected temperatures on different components. The surface temperature of our hardware varied by almost five hundred degrees Fahrenheit depending on the orbit and whether it was in the sun or darkness. That is why we'd buried hundreds of temperature sensors in the ISS hardware to provide insight into the health of the systems. These were monitored closely because unusual temperatures could indicate that the hardware was not performing as expected or could cause stresses that would reduce hardware lifetimes. We saw odd temperatures on several key elements, but as the holidays approached, nothing was serious enough to demand action.

Throughout most of the year, the station passes through sixteen roughly equal periods of sunlight and darkness each Earth day. This provides nearly equal periods of warming from the sun on the day side and cooling on the night side of each orbit. However, due to the relatively high inclination of the ISS orbit, around the time of the winter and summer solstices, the station remains in sunlight continuously for several days. The completed station was designed to accommodate this through normal operations, but the early station had to keep temperatures in balance by periodically changing its attitude to present more or less of its surface to the sun.

That holiday season was our first chance to see how our new hardware would tolerate this extreme thermal environment, and the temperatures we observed did not align with the preflight analysis. Some boxes were getting too hot, while others were getting too cold. Worse, some of the attitudes we were using to manage the temperatures were causing the Soyuz capsule to overheat. This would cause its fuel to heat up and eventually deteriorate, which was unacceptable because the Soyuz was the crew's ride home, and we had to take care of it.

The MER was buzzing around the clock with experts analyzing the data and reviewing their findings. Most temperatures were within or close to the expected range, but a critical few were not. Reorienting the station daily did not keep them from drifting out of limits. We were particularly concerned about an MDM—a computer used during early assembly—and the KU gimbal, a joint needed to point the high-data-rate communication antenna at its ground station.

Tommy heard rumblings that something was going on and tracked me down.

"We're having real problems with thermal management," I told him. "In order to protect the Soyuz, we're spending time in attitudes that make the MDM too hot and the KU gimbal way too cold."

"What does that mean?" he asked.

"We can manage most of the temperatures, but not all of them. The Soyuz is obviously our top priority and must be protected at the expense of other things. The lifetime of the MDM will be shortened, but we won't need it much longer. When the KU gets too cold for the motor to drive it, we'll lose video and high-rate data, but it'll come back once we warm it up. It's mostly a problem for public affairs events, so we can live with it. The big problem is not a matter of whether we'll exceed specifications on any boxes but which ones we choose to push past their limits."

Holloway sat back in his chair and looked reflective.

"What does this mean for the future? Are things going to get worse?"

Orbital Inclination

Inclination is the angle the orbit makes with respect to the equator and roughly equals the latitude at which the object is launched. Shuttles launched from the Kennedy Space Center at latitude 28° while Russian launches occur from Baikonur at 51.6°. This required the ISS to be at 51.6°, so all of our assembly flights had to launch at a higher inclination. Because eastward launches close to the equator benefit from the rotational speed of the Earth, going to the higher inclination to accommodate our Russian partners reduced the mass a shuttle could launch to orbit by about ten thousand pounds. Shuttle modifications eventually recovered most of that loss, but launch performance remained a challenge to our planning.

Tommy's demeanor shifted subtly. No longer acting as a senior manager, he was reverting to his former role as a flight director. His technical instincts were kicking in, and I hoped this meant he would finally start to appreciate the challenges I faced every day.

"I think it's just an artifact of our early assembly configuration and the fact that we're flying it through the solstice," I said. "Some of the equipment is only needed for a short time and doesn't have to last forever, while some is in temporary locations and should be better off later. The problem is made worse because some equipment is being worked harder during this early stage of assembly. All in all, it should get better soon, but it will require a lot of care to get through this challenging time."

He nodded and left.

Meanwhile, the plans to take the family to Michigan were approaching a decision point. We were either going to leave soon or give up on it. Doris and the kids were eager to go, but I was concerned enough that I wanted to stay. I could keep in touch with the team by phone and email, but I preferred to discuss such serious issues face-to-face while reviewing the data with my experts.

As it turned out, a major ice storm came through that made the trip in our small plane impossible. Driving would be a two-day ordeal each way, and none of us wanted that. We quickly bought a tree and made plans for a holiday at home.

Long before dawn on Christmas morning, I called the control center. The MER manager reported that hardware temperatures were responding as expected, and the flight director said that the crew was content. I left word to call me if needed and that I would be in later.

The morning with my family was a delight. I lit a fire in the fireplace, Doris made us a special breakfast, and I put on a CD of my favorite instrumental music as we opened our presents. Then, at Doris's encouragement, we sat by the dancing fire as I told the kids the full story of Phase 1 so they could understand something about the important work that had taken me away from them so often. Despite their disappointment about the trip, the kids enjoyed the holiday. The occasion reminded me of how much we were all sacrificing for ISS and deepened my appreciation for our precious time together.

I drove to the center a little after noon. Mine was the only car in sight as I approached the guard, and he returned my warm greeting. Although dozens of cars were in the parking lot, there were no signs of activity as I made my way to my spot and entered through the security checkpoint.

The cavernous MER presented a discordant mix of Spartan government facilities and Christmas cheer. Every open surface was decorated with family pictures, garlands, and holiday displays. Cookies and other treats were plentiful, and a lovely sense of community filled the air. I felt privileged to share this time with these great folks and did my best to thank them for their dedication. My years sitting on air defense alert had taught me about the pain of being away from family on important holidays, and I was heartened to see such a bright and positive atmosphere. People were proud to be there and happy to share the experience with their colleagues. We were a family, too.

The flight control room was much the same, though the group was smaller. Cookies and other treats were spread out on a table near the entrance, while family pictures and other reminders of the holiday were prominently displayed at workstations. The room felt friendly and relaxed.

I requested a private communications loop to call Shep and see how he was doing. He sounded happy to hear from me and was clearly enjoying his time in space. He mentioned there were some things we should talk about, but not on the holiday. I agreed to get back to him later in the week.

The next day, things were a little more normal. Although most of the team could have taken the holiday period off, many of the NASA and Boeing teams were busy preparing for the next major event: the launch of the largest American module—the aptly named science laboratory *Destiny*.

28. MEETING OUR DESTINY

The new year began with a strong sense of optimism and purpose. More than at any other time in my life, the time, place, and job all felt exactly right, which gave me the confidence to face the challenges ahead. I knew I would need it during this time, the most critical year in the ISS assembly. In 2001, we would install and activate its core systems in a series of incredibly intricate assembly missions. We had planned for a quiet December to give the team a chance to prepare for the new year's challenges. Despite the frustrating thermal issues, most people had gotten the rest they'd needed.

The team was psyched and ready to go, and it felt good. Everyone knew the pressure was real—the race to get hardware up and operating before a failure put us in extremis. We had accepted and planned for this risk, so we faced it with calm and confidence.

The *Destiny* laboratory would be the focus of the US segment. Its design was traced to Space Station Freedom, and many of us who had been wrenched by the demise of Freedom felt a special gratification in seeing it approach launch.

Because *Destiny* would be located near the station's center of gravity, it provided the best research environment and would hold the majority of the US research facilities. To take advantage of its prime location, it was configured with extra power and cooling. The US research community had been waiting a long time for these facilities to be in place.

But the lab was also the key to the station's early functions. It carried many critical systems, including computer controls, life support, and thermal control. Its extreme complexity, with thousands of design and test requirements, caused it to take about two years longer than planned. Part of this delay traced back to the early days of the program when ground testing, including integrated testing right before launch, had been cut to save money. Wiser heads ultimately prevailed, and integrated testing was reinstated. It successfully flushed out many problems that could have become crises on orbit. Officially, NASA blamed the ISS assembly delays on the Russian schedule for the *Zvezda* module, but we knew that if *Zvezda* had been on time, *Destiny* would have caused the same delay.

As the launch approached, I heard that a member of the shuttle crew was expressing concern about installing the lab. Astronaut Marsha Ivins would use the shuttle's robotic arm to extract the lab from the cargo bay and berth it to the forward port of *Unity*. She reportedly said that if there was a problem installing *Destiny* on the station, she would not even try to put it back in the cargo bay but would jettison it instead.

At first, I ignored the report. The crew did not have the authority to make such decisions except in the direst circumstances. I knew Marsha pretty well, including her well-earned reputation for making disparaging comments about engineers and managers. I imagined she was just waxing eloquently about how she was prepared to save the day from us idiots in the program office. But the rumors persisted and began to raise concerns with other members of the ISS team, even reaching Holloway. Although I thought the risk of a problem was small, the stakes made it prudent to determine whether Marsha had legitimate concerns. I asked to attend her next training session to understand the potential problem, but Marsha said she would rather set up a separate session to talk it through. We met in the Shuttle Engineering Simulator in Building 16 so she could demonstrate the challenges of removing and returning the lab from the cargo bay.

The Shuttle Remote Manipulator System (SRMS), also known as the Canadarm because it was built by the Canadian Space Agency, was a key feature of the shuttle, used to deploy and retrieve objects in space. Although it had been used with many satellites, it was coming into its own as a critical tool for ISS assembly. A later flight would deliver its younger sibling, the Space Station Remote Manipulator System (SSRMS), to the ISS, where it would live and work for the life of the station.

Both arms were exquisite contributions from our neighbors to the north. The SRMS had a shoulder joint attached to the shuttle, an elbow joint for reaching, and a wrist joint and end effector for grabbing and manipulating hardware. When it worked as designed, the operator controlled only the position of the end effector; the joint angles were calculated and driven by intelligence in the system. If there were problems, the operator had the option of controlling each joint separately. The SSRMS was similar but had an extra joint and redundant hardware for reliability. It also had an end effector on both ends, allowing it to walk itself around the completed space station.

Marsha was a very experienced arm operator; the normal removing of the lab from the cargo bay and putting it back, if necessary, was easy for her. But our conversation showed that she worried that if she had to return the lab to the cargo bay using a partially failed arm, she might damage either the lab or the shuttle. I was sympathetic, but I also knew that this risk was extremely modest compared to many others in the program. I thanked her for sharing the challenges she faced with me and assured her we would take them into account in the unlikely event she was asked to perform that task. After our talk, the rumors about jettisoning the lab ceased.

The shuttle team did their usual superb job with the launch, and *Destiny* was soon on its way to its new home. It would take two days to reach the station, leaving plenty of time for the shuttle crew to get set up and ready for the busy docked period.

When I arrived in the Mission Evaluation Room early the day after launch, I found the engineers discussing a potential problem with some heaters. The laboratory had five sets of electric heaters on its hull to keep it warm enough to protect the water in its coolant loops. Thermostats on the hull cycled the heaters on and off to maintain the correct temperature, but an early cost-cutting measure had deleted the sensors that would allow us to monitor the lab's internal temperature and air pressure. Instead, we had to monitor the electric current flowing into the heaters to judge whether things were within limits.

The preflight calculations indicated that the heaters would run continuously for about sixteen hours, after which the thermostats would begin to cycle on and off. By the time I arrived, the heaters should have started cycling, but they were still running at full power. The team had already reviewed the preflight analysis and found it to be correct. After a quick review, we agreed the best course was to wait a while to see if the heaters would start to cycle.

Two hours after the heaters should have turned off, concerns grew. After four hours, the engineers were experimenting with new analyses and different assumptions. By the time six hours had passed with the heaters still running, the engineers felt the most likely cause was a large hole in the insulation that was allowing heat to radiate into the cargo bay.

The heaters were the main issue at our engineering review before the MMT meeting. We reviewed every aspect of the heater system and its operation, along with the potential failures that could have caused the problem. The MER engineers explained that each of the five independent heaters on the laboratory hull was controlled by its own solid-state thermostat. All five heaters were still running at full power, and it seemed very unlikely that all five thermostats would have failed simultaneously. If the thermostats had indeed all failed in the on position, the temperature inside the lab would be approaching 115 degrees Fahrenheit. Although that temperature was nowhere near high enough to damage hardware, as the air in the lab warmed, its expansion would increase the lab's internal air pressure until it triggered a pressure relief safety valve that would vent some air to prevent structural damage. If the valve did open, there was always a chance that some dirt floating in the module would cause a leak and allow all the air to flow out. We could recover from such an event, but it would be the kind of dire event that would slow the hectic assembly sequence we had all spent so much effort and worry to keep on track.

After an hour of intense discussion, the engineering meeting ended with no consensus as the Mission Management Team members filed in to decide on what to do. While we argued through the options, the room filled with anxiety. The MER team recommended that the power be cut off for the heaters, while MOD argued against it because of the possibility that whatever failure had caused the heaters to stay on would not allow them to be switched back on later. Polling the membership of the MMT did not resolve the issue. Most people said they could support either answer. As was so often the case, the decision rested with the chair—me.

I decided to shut off the power and asked MOD to be prepared to manage the temperature with attitude changes if necessary. With a clear decision, the room suddenly became calm. All the emotions and arguments were forgotten, and the team went off to execute like consummate space professionals.

The docking and installation of the *Destiny* lab proceeded as planned. When we finally got the module's computer powered up, we found that all the thermostats had indeed failed despite extensive preflight testing. The air temperature inside had reached almost 120 degrees Fahrenheit, but the overpressure relief valve had operated as designed, venting enough air to keep the pressure within limits, and no dirt had jammed the valve. After three days of high anxiety, all was well.

The mission also included three spacewalks to activate the lab and prepare for upcoming missions. With the lab and its systems activated, the control moment gyros (CMGs) on the Z1 truss began providing attitude control, reducing the amount of propellant needed by the Zvezda module's control jets. We also got some optional tasks done to get ahead for later missions in case problems arose downstream.

One of the most anticipated features of the new module soon led to a string of headaches that would continue for years. The Window Observational Research Facility (WORF) was a jewel of the early research program because it could be used even during the early assembly period when the crew had little time or material for research. Cameras of different kinds were set up in front of the window and operated remotely from the ground to conduct science. To maximize its scientific value, the window was made of extremely high-quality optical glass and carefully mapped to document all the minor distortions that remained.

To ensure its research value, the window had to be protected from contamination by visiting vehicles, especially the shuttle. The window faced directly into the shuttle's cargo bay during the early docked missions, exposing it to the gases given off by the materials that lined the bay. We asked the crew to keep the window covered except when needed, but the astronauts could not resist looking out of the station's best window.

The issue was brought to my attention by astronaut Mario Runco. He and the science team soon convinced me that we needed a flight rule to keep the WORF covered whenever there was a risk of contamination, including most of the time when the shuttle was docked. Since the cover required only seconds to open or close, it seemed trivially easy. I directed MOD to observe these guidelines while we pushed through a formal rule.

But the crews loved the view, particularly when the shuttle was approaching and docked. Repeated requests to keep it closed were insufficient. I pushed MOD to expedite the formal flight rule, but even after the rule was published, the problem persisted.

The space program is populated with some of the smartest and most energetic people in the world. Such smart and aggressive people sometimes believe that any decision they do not agree with must be the product of a stupid bureaucracy and unworthy of their attention. Those who did not respect the concern about contaminating the window decided to ignore it as they saw fit.

I soon lost patience with their cavalier behavior. One of the key reasons ISS had survived the challenges of its history was the support of the science community, so the station program needed to respect that community's needs as a reward for their support. The WORF offered researchers around the world a high-quality vantage point for observing the Earth at a time when we could conduct little other research. All we had to do was take care of it, yet too many were unwilling to do even that. It was time to enforce some discipline.

NASA television broadcasted live images almost continuously during shuttle missions, including many lovely external views taken by cameras in the cargo bay. It now became a habit that anytime I saw a picture of the lab, I checked the window. Far too often, I found it open in defiance of the rule. In the beginning, I would call the flight director and ask why it was open before requesting that it be closed, but before long, I became more forceful.

It even followed me home. One evening, I sat in the family room, eating my late dinner while watching NASA TV. When the on-screen view showed the outside of *Destiny*, I once again saw the window open. I picked up the phone and called the flight director console.

Without even introducing myself, I growled, "Close the damn window!"

The flight director, one of those who did not much care for managers or science rules, did not argue. He simply said, "Got it."

About a minute later, I heard Capcom tell the crew, "We'd like you to close the shade on the WORF." Moments later, it swung closed. My teenage son, playing on the family computer nearby, grinned broadly!

Eventually, the word got out that this rule was for real—a message reinforced a year later when the science team received the prestigious Stellar Award for having fought for it.

The increasing pace of operations was exhilarating, but it challenged us to stay ahead of what was going on. People began to describe the pace of a shuttle flight as a sprint and ISS as a marathon. The shuttle team could slow down for a few weeks between flights, but the core ISS team was on duty continuously. Its experts were on call all day, every day.

The pace was intense. Including Russian launches, we were averaging a launch every three weeks, with numerous critical mission events—whether installing modules or activating systems—every week. The reviews needed to plan this activity came in a torrent, keeping people locked in meetings and telecons almost constantly. Because of the time difference with Russia, many telecons began in the wee hours of the morning in Houston, adding to the burden. My typical day included six to eight hours in the control center, followed by a similar amount of time in planning and design reviews. Weekends and holidays passed almost without notice.

Some critics of ISS complained that we were putting schedule

> **Reboost**
>
> Although we talk about things in orbit being in a vacuum, the atmosphere actually extends upward far beyond the ISS orbit, becoming thinner the higher you go. At the altitude the space station orbits, there is sufficient aerodynamic drag to lower the orbit by about one hundred meters per day. In order for the station to achieve its planned lifetime, this altitude must be made up by periodically reboosting to a higher orbit. The shuttle often did this, but we could also use the rocket engines on the Russian elements. The propellant is kept in reserve for this reason and to provide attitude control should the CMGs fail. Conserving precious propellant was the reason for activating the CMGs as early as possible.

above safety, and I often found myself explaining that the aggressive schedule was actually driven by safety—the need to build the station's capabilities before some critical failure occurred without enough redundancy and margin to manage it. We were not rushing; we were executing the plan we had built. Most importantly, we had staffed appropriately, and the working troops were doing fine. They were excited to finally see the station coming together and did not want to miss a single hour of activity.

The activation of *Destiny*'s systems was a magnificent achievement. Every day gave us more margin for dealing with potential problems that would someday come.

29. BUILDING MOMENTUM

A s February arrived, positive energy continued to surge through the team. More than fifteen years after President Reagan had announced that we would build "a permanently manned space station," we were doing it. The successful activation of the *Destiny* lab triggered celebrations and positive coverage in the media. Our schedule was chockablock with upcoming missions, and the Space Station Processing Facility at the Kennedy Space Center was filled with modules being prepared for launch. Its main hall was a beehive of activity, and every visit showed reassuring progress on our plan.

The program was finally gaining broader agency-wide involvement. Morale and interest soared across NASA, its supporting industries, and literally around the world as people saw their handiwork in a steady stream of enthusiastic media stories.

I certainly felt the ambient energy, but when I paused to try and savor our success, I could not help thinking, *So far, so good.* A big part of my job was to worry about all the ways things could go wrong and be sure we were prepared for them.

While most of the station's systems were working beautifully, a few things had started to glitch. The issues had begun on the Russian segment that was launched first, but soon enough, our American systems started to show similar behaviors.

For example, the Russian carbon dioxide removal system launched in *Zvezda*, and it immediately showed signs that some of the filter material dislodged during launch and jammed a valve. This did not cause it to fail, but it did cause a minor leak that reduced its efficiency. Jay Greene quickly attacked the Russian hardware as primitive and poorly made. A few months later, we activated the equivalent US system in the *Destiny*. NASA's system used the same technology, though built by American vendors, and most of the NASA team thought such a failure was virtually impossible due to our outstanding design. I had to smile when the American system had exhibited an almost identical failure upon activation.

This was another of the many opportunities where we might have gained from the Russian experience, but ego and chauvinism had blinded us. Phase 1 had provided similar lessons, but very few people outside of our tiny team had paid attention. When I had introduced these lessons to ISS meetings, many had not considered them relevant. This hubris had caused us to suffer a number of avoidable, though mostly minor, problems that our Russian colleagues had already experienced.

The workload continued to grow. More hardware in orbit meant more data to analyze and more anomalies to understand and address. More missions and events meant more stories for public affairs to handle. New people were joining the team, and those with more experience were moving up to become leaders. We were at the center of the most profound team experience I had ever encountered, and it was a joy to behold.

As always, my days began early with a detailed team review of the anomalies being encountered, and everyone worked hard to ensure that we maintained NASA's high standards despite the workload. Nobody was willing to let minor deviations slide by. The engineers, contractors, and vendors went far beyond formal requirements to support us. Their pride and enthusiasm were infectious.

Even the media seemed to appreciate what we were doing and why. Some of our regular reporters were our best spokespeople, becoming very knowledgeable and supportive. Their professional and well-informed reporting kept the inevitable small group of naysayers from making a big deal out of every little surprise.

Some unanticipated problems started small but grew to be concerns, such as when a shuttle crew decided they liked the Russian toilet on the ISS more than the American one on the shuttle. The increased usage quickly consumed the full supply of a critical filter, leaving the station crew in a precarious position. The solution was to modify a shuttle toilet filter for use in the Russian unit. This required the Russians to share their design data and NASA to experiment with our filters to make them fit. The complete success of the exercise showcased the kind of technical and human cooperation I had been working toward since the early days of Phase 1.

Engineers pored over system performance data to compare predicted behavior with reality. Some issues they found seemed trivial but turned out to be important. A few materials on the outside of the station that had tested fine on the ground darkened in color after a few months on orbit. The color change was caused by exposure to trace amounts of gases given off by other hardware. The darkening affected how much heat the materials absorbed or reflected onto other surfaces, which altered the delicate thermal balance of the space station.

The enormous cost of fixing things once they got to space meant we needed to make adjustments as soon as possible. As soon as we saw adverse trends, we looked for

Ground Testing

The extensive testing program for space hardware is limited by cost, schedule, and our ability to simulate the space environment on the ground. Engineers have developed techniques for simulating the radiation, vibration, and most thermal loads. However, some things, like convection, the gravity-driven flow of gas caused by density changes with temperature, could not be removed on the ground, and available zero-gravity tests in aircraft did not last long enough to show the problem. We had to find these problems once the hardware was in space.

ways to protect the hardware already launched and improve the hardware still on the ground.

Although everyone tried to find and fix errors during ground testing, a few got through. Some could not be detected because of gravity or other effects impossible to simulate on the ground, but others slipped by because testing was either not done or cut short for cost or schedule reasons. Those were especially embarrassing.

Cost is a constant nemesis for space systems, and programs are always under pressure to save money. Sometimes, this leads to the use of an existing component in a role for which it was not intended rather than spending time and money to design, build, and thoroughly test a new one. This can result in barely adequate solutions being accepted because they appear to be "close enough."

One of the places we saw this was with the so-called beta joints that allowed the solar arrays to track the sun as the station orbited the Earth. In the original Freedom design and the final configuration of ISS, with the arrays positioned on the truss to the side of the station core, these joints only had to oscillate back and forth a few tens of degrees. However, in our early ISS configuration, while they were on top of the Z1 truss, they tracked the sun by rotating 360 degrees around their long axis, pointing at the sun even when it was behind the Earth. That meant they rotated continuously in the same direction. Engineering analysis concluded that the joint, originally intended for oscillation, should work fine in a continuous motion.

That analysis proved wrong in an exceedingly subtle way. Within weeks of activating the joints, we started to see the current drawn by the motor slowly increase, indicating that it was working harder for some reason. The most likely cause was increased drag in the joints, but we could not imagine how this could be. The load on the bearings was incredibly small, and analysis suggested they should work well for many years. However, the problem was real, and although small at first, it eventually became significant enough to stall the motor and stop the array from moving.

After weeks of study, the best guess for the cause was that continuously driving in the same direction was causing the individual bearings in the race to bunch together behind the one with the highest drag. This would not happen on Earth because gravity loads would keep separating the ball bearings as the mechanism turned. The proposed solution was a

> ## Workarounds
> When problems arise, the most expedient solution is often to operate the hardware differently than it was designed. Such steps are called "workarounds" because they are designed to compensate for things not working as planned. Some workarounds are temporary, while others could be used for years.

procedural workaround: periodically turning the joint in the opposite direction. If our guess was correct, this would separate the bearings within the joint and reduce their drag. The fix was partially effective and usually—but not always—brought the currents down for days or weeks at a time. The inconsistency left us worried about how long the joints would last.

Backdriving the joint seemed to be working often enough and well enough to allow us to get through the assembly period, but if we were wrong, we stood to lose most of the power for the American side. Because the bearing's behavior was unpredictable, I felt it was imperative to be prepared to fix it. We had spares on the ground, and the joints were designed to be replaced on orbit, but it would take three challenging EVAs to make the changeout and significant replanning of the shuttle manifest to carry a replacement joint. Training the crew to do the EVAs would take long enough that I wanted to start right away.

After discussing the issue with the MMT, I directed that the replacement joint be ready to add to an upcoming mission if required. The Space Shuttle program was also asked to develop plans to accommodate the activity on one or more of the upcoming missions. These decisions drew only minor grumbling, but when I asked to have the final procedures prepared and a specific shuttle crew undergo initial training, both the Space Shuttle program and the crew pushed back. They did not want to do additional work for something that might not be required.

I understood their reluctance. I had devoted a great deal of time and energy to understanding the workload on the crew and controllers and had weighed that against the criticality of a possible failure when I'd made my decision. Still, the crew pressed their objection to Tommy Holloway, who called me into his office to discuss the complaint and review my reasons for pushing ahead.

After listening to my explanation, Tommy decided the impact was great enough that he wanted it reviewed by the Space Station Control Board. This board had authority over any decision made anywhere in the program, although, to this point, it had not scrutinized what I was doing through the Mission Management Team. This was mostly because all the organizations on the control board were also members of the MMT and were represented there by technical experts. But Holloway and the other managers did not usually attend the MMT meetings, so they wanted to have their say in a forum senior to mine.

The presentation was quickly added to the next control board meeting, and I gave the briefing personally since it had been my logic and judgment that had driven the decisions. Some of the managers used the opportunity to ask difficult and penetrating questions, which I was glad to address. Our ready answers clearly showed we had the issue under control.

After a long discussion, Tommy polled the board on whether to accept the MMT decision. The decision was unanimous to proceed, eliciting strong support for the rigor and care we had exercised. This was the only time the control board reviewed a decision I made through the MMT. Fortunately, the workaround continued to take care of the problem until the solar arrays moved to their permanent home on the truss, so we did not have to execute this difficult repair.

Other less serious in-flight anomalies arose constantly. We tracked IFAs in the back of the MER on a large board that showed several hundred in total, with about half already closed.

In addition to tracking our progress, I saw this list as an indicator of the stress level on the team. Spirits remained high, and people remained eager to do the work, but they were spending very long hours with no end in sight. While I worried about their health and family situations, my bigger concern was the quality of people's judgment and attention to detail. Fatigue makes people more likely to miss important nuggets in the heavy flow of data. I began to spend more time looking for signs of trouble, not only in the hardware and software but also in the team. Some people thought this meant I did not trust them. Others felt that their success ought to allow them the autonomy to make final decisions on their own. A few even resented the way I asked about the logic behind their recommendations—especially the way I required final concurrence through the MMT or another review board.

In fact, I was extremely pleased and proud of the team. They cooperated brilliantly and did outstanding work. But success was no reason to let down our guard, and that high morale itself could cause problems. If it overinflated their confidence, that could also affect their judgment and lead to mistakes.

I was particularly concerned about the flight directors, especially the newer ones who'd been hired and trained specifically for ISS assembly. Most of them had spent years preparing for the missions they were leading, and they were enjoying the work enormously. Some of them could be found in the control center around the clock, watching and sometimes meddling in operations that were not their responsibility. I usually just watched and left it to MOD to manage them, but occasionally, I had to speak up.

A couple of flight directors became so consumed by the work that they started to lose touch with the rules and boundaries of their authority. Others made excuses to avoid coming to the MMT meetings and sent another flight director instead. Although this was sometimes necessary, MMT meetings dealt with countless details that benefited from discussion with the whole team. Our fragile working relationships across organizations required that everyone understand and concur on the resolutions.

One of the newest flight directors seemed to be diverging from the norms. I had known him for years and had seen him do great things. Although I respected and trusted him, he was withdrawing from the management process and becoming increasingly bold with his own decision-making. Whenever I stopped in to see him, he was friendly and respectful, but I noticed that he either would not mention the little changes he was making or would trivialize them. He seemed to be taking the path of "better to ask forgiveness than permission." These little things made me worry that he was cutting corners elsewhere—seriously bending the flight rules. I asked my team to give special scrutiny to the way he followed them.

Before long, I got reports that he was indeed stretching his authority in ways I had not seen. None of them were a big deal on their own, but in such a tightly integrated program, their consequences could cascade into something serious. I needed to get him back in the box.

When I discussed it with Randy Stone, he was disappointed but not altogether surprised. He immediately took full responsibility for fixing it. Once again, he put a senior flight director on the MOD management console during major activities to review and oversee the person's actions. It took some finesse on Randy's part to ensure that someone was present enough to be effective but not so obvious that the flight director felt he was being watched. They did a great job, and I do not believe the individual ever figured out that he was the subject of such attention.

I also discovered a few of my own people skirting our procedures to give crews things they wanted—often little comfort items they felt should not need or would not receive formal approval like Shep's power saw. Our processes try to accommodate crew desires, but sending things to space is not like packing for a camping trip. Every ounce added to a manifest comes at the expense of some other ounce, and each item has to be thoroughly scrutinized for any possible harm to the crew or spacecraft.

Some hazards are obvious, like flammability and incompatibility with zero-g. Others are less so, such as possible dangers to the crew or systems if the item broke free and went floating around the spacecraft. Some risks are subtle—like a tendency to shed crumbs, dust, or shards that, in the absence of gravity, float in the air and create a hazard to breathing and eyesight. Some things that are fine for the American system can disrupt the Russian system. The stowage of small items left on board can accumulate into a serious problem over the years.

Crew time was as tightly constrained as upmass. When my people winked at little unapproved changes to the schedule, they upset our carefully negotiated balance of tasks. In one case, an unapproved change ended up slighting some research that NASA had spent tens of millions of dollars to include. In another, the crew wanted to forgo their scheduled exercise for some preferred activity, which would be detrimental to their health in the long term.

My job included the unenviable role of enforcing our rules and procedures. I knew some people considered me a hard-ass and a mindless bureaucrat, but I was not. I considered every request as fairly and generously as I could, but not all could be accommodated. I admit the criticism stung, but like other aspects of this job that "nobody else wanted," it had to be done.

30. STORM CLOUDS

By the beginning of March 2001, my days had settled into a rhythm that made the hard work more manageable, though it did not free up any more time for my family. I was always gone before they awoke and usually got home as the kids were on their way to bed. My dinners waited in the microwave, and my wife and I squeezed our quality time together into the length of a walk around the neighborhood—often after dark, always holding hands. Although I knew I was neglecting my duties to them, their tolerance and the real progress of our important work kept guilt at bay. I silently hoped for my family's patience and forgiveness for as long as it took.

That important work continued with Flight 5A.1, the first of what we hoped would become routine deliveries of logistics to the ISS. Earlier flights had carried logistics, but this was the first use of our dedicated logistics carrier, the Multipurpose Logistics Module (MPLM), which was part of the European Space Agency's contribution to the international program. The MPLM berthed to the *Unity* node.

Flight 5A.1 also exchanged station crews for the first time, and with that exchange, Expedition 2 took an important next step in the maturation of our international cooperation. Where the Expedition 1 crew had been led by an American commander, Shep, and had had two Russian flight engineers, Sergei Krikalev and Yuri Gidzenko, this crew had a Russian commander, Yuri Usachev, and two American engineers, Susan Helms and Jim Voss. This alternating of command responsibility was important to maintain the trust and respect between our teams.

Although ISS now had a Russian commander, the leadership of the mission remained with NASA. I, as the chair of the MMT, retained final authority and responsibility for all operational decisions. For the first time ever, a spacecraft commander of one nationality would be taking direction from the ground control of another. I resolved to ensure he would not be forced to decide between taking direction from Houston or Moscow.

Usachev was an excellent choice for the first Russian commander. An experienced cosmonaut, he had already shown his willingness to work with Americans as one of the two cosmonauts who had gotten along so well with Shannon Lucid on Mir. Although I had met him, we did not have a personal relationship to help us through any difficult situations. I had always actively sought Russian input for significant decisions—they participated on the MMT as full members—but with this precedent-setting mission, I redoubled my efforts to resolve conflicts between the ground teams to spare Usachev a difficult choice.

Unloading the MPLM was among the early activities of the shuttle mission. After entering the module, they sent down compliments to the KSC team on how clean the module was and huge kudos to the ground processing team. Much of the

> **International Participation**
>
> Although most of the events outlined in this book involve only the US and Russia, the ISS was always a truly international effort. The US contributed many elements and managed the work. Canada provided the robot arm, a dexterous robot for performing maintenance, and other logistical elements. The European Space Agency contributed its own laboratory, the MPLM, and other elements, while Japan contributed a laboratory, logistics module, and other elements. Although I participated in the planning and integration of those elements, my time was mostly taken up with the hardware on orbit and keeping the Russian-American team working together.

dirt tracked into the module during loading, along with tiny shards of metal left behind from manufacturing or generated by wear, was almost impossible to find while on the ground, especially once the module was stuffed with cargo. In space, this tiny debris would blossom into view as it floated free in the air. This was why the cargo unloading procedure called for the crew to wear eye and respiratory protection in a new module. Using protective equipment is always awkward, and some crews stopped wearing it if they thought there was no obvious risk. The cleanliness of the MPLM caused the crew to follow that lead.

The MPLM was designed for two kinds of racks. Some held complete systems or experiments to be moved to the ISS intact, while others held stacks of cargo bags that would be unloaded one at a time. The bags had been created for the Mir program and provided superb loading flexibility, facilitating the late changes that had become a fact of life. However, while we'd had lots of experience using the bags in the Spacehab module, this was the first time they would be used with these stowage racks.

When the first close-fitting cargo bag was pulled out of its slot, the suction behind it drew a small cloud of aluminum chips and dust into the aisle, aimed right at the face of the crewmember doing the work. One of these shards got into an eye, causing work to stop while the injury was treated and dirt was removed from the air. The eye healed without difficulty, but it was a close call that reminded everyone why the requirement for protective gear was in place.

The rest of the mission went well overall, though there were plenty of things to keep everyone busy. A smoke detector triggered in the *Destiny* module, causing the use of an emergency procedure and a search for the cause. They found no sign of fire or smoke, suggesting the smoke detector had been triggered by a small cloud of dust, as had been common in the early days of the Space Shuttle program. The extensive procedures used all throughout space missions had their origin in similar problems seen in the past.

Expedition 1 ended with a picture-perfect entry and landing. I was eager to talk with Shep about our new station, but a long list of doctors and others wanted his time first. In the meantime, I looked forward to learning all I could from the shuttle crew.

To my surprise, the day after landing, Shuttle Commander Jim Wetherbee came to see me. He and I shared a concern about people becoming complacent, and he wanted to tell me privately about some events that had occurred during his mission. One of his crew had ignored an important protocol and had had a close call that could have caused a serious accident, while another had not followed a checklist in a way that might have killed a crewmember. Nothing bad had actually happened, but Jim was irate that two veteran astronauts had foolishly put the mission at risk. We discussed the specific actions of each and what they could mean in the future. He assured me that he would work with the astronaut office to keep similar mistakes from happening again.

Meanwhile, the Russians had been trying to sell open seats on Soyuz missions to paying passengers and had identified the American millionaire Dennis Tito as the first space tourist. In the past, the only visitors to the Mir had been "guest cosmonauts" from friendly countries. With ISS, Russia offered rides to anyone who could pay their high price. Tito would launch immediately after the next shuttle flight left the ISS. We all expected commercial flights to happen someday, but many were concerned that this was happening too soon.

My concern was that a poorly trained visitor could inadvertently upset our finely tuned plans. For example, if they inadvertently turned off a system, it could take hours of work to restore. Both NASA and Russia had experienced situations where such problems had put the mission at risk. We'd had recent incidents in which a fully trained, professional astronaut had accidentally vented some valuable air out of ISS and a cosmonaut had caused a serious spill of carbon monoxide on the Mir.

When asked my opinion, I agreed to accept Tito as long as he was strictly supervised and warned to keep away from the critical station systems. In truth, I was less worried about Tito than about the financial stress that made his mission so important to the Russians. The history of both the American and Russian space programs had many examples of how small judgments driven by cost or schedule had caused major problems.

Eight years earlier, when I had first briefed Dan Goldin on the risks of including the Russians in our space station plans, I had identified the government's chaos as the biggest unknown. The risk from the poor Russian economy remained; if Tito's flight could take some pressure off the Russian Space Agency's budget, I thought it was worth it. Many of my colleagues disagreed, and some fought the issue as hard as they could. In the end, we had no control over the decision.

I would have to find a way to manage whatever came of it.

31. CRISIS COMES

One of my favorite Space Station Freedom memories was a conversation with John Aaron about his days working on Skylab, NASA's first space station, back in the 1970s. John said that, for him, the transition from the fast-paced, high-stakes environment of Apollo to the more relaxed Skylab had been the difference between a challenging job and utter boredom. He admitted that the early problems Skylab had experienced after it had been damaged during launch—a section of the rocket's outer shell had ripped off, taking a solar array and insulation with it—had given the team a month or so of excitement. But when that had been over, all he could see in the future was unrelenting tedium. He'd left the program as quickly as he could.

I certainly did not see such tedium in our future. The ISS roller coaster had reached full speed, and though we couldn't anticipate the dips and turns ahead, the program's aggressive plans promised many years of challenge and excitement for those of us taking the ride.

The few weeks until the next shuttle flight filled up with the work of activating and checking out the equipment in our new *Destiny* lab. I could not help noticing that our relations with the Russians seemed smoother while we Americans were preoccupied with our hardware and out of their hair. I am sure they noticed this as well.

The next assembly flight, called 6A, would bring another MPLM to outfit *Destiny*, along with one of the most critical tools needed for the rest of assembly—the Space Station Remote Manipulator System (SSRMS). SSRMS was an advanced version of the shuttle's SRMS arm that had flown many times, and it was the core of the robotic maintenance system that was Canada's contribution to the international project. We needed the new arm, as the station would soon become too large for the shuttle arm to reach worksites. Longer and more capable than its shuttle predecessor, it could move to where it was needed by attaching itself to specialized mating ports located around the station.

After another flawless launch and rendezvous, the new MPLM was berthed and its hatches opened, and cargo transfers began. The SSRMS and its carrier were soon moving from the shuttle to the ISS. Sitting with my increment and launch package teams, I watched the operations team work its magic.

The calm was interrupted by a call from one of the flight controllers: The Command and Control (C&C) computer on ISS had failed. We expected this would be only a minor inconvenience, as the station had three C&C computers. The team would have the backup operational within minutes.

The C&C computer has an executive role in managing ISS systems. For example, it tells the computer in the life support system what temperature to maintain in the various modules rather than controlling the temperature itself. The C&C computer is not needed for moment-to-moment control, but it is critical for readjusting systems during times of transition. Its commands are infrequent but extremely important.

Most critically, the C&C computer manages the attitude control system, which works differently depending on the station's configuration. For example, when the shuttle is docked, the ISS's center of gravity is very different from when it is not, requiring different gain settings to stabilize and maneuver the combination. Without the right settings, the control system could be overworked, and propellant could be depleted quickly. While things were okay for the moment, we needed the C&C computer to work before the shuttle undocked.

That undocking was days away, and we had plenty of time to restore the lost function. With no urgent need for it, the flight director sought to restore the failed computer rather than switch immediately to the backup. When they could not coax it back to life, they turned to the backup computer and brought the cold backup to hot standby.

The backup computer responded to commands but could not communicate with the rest of the system, and the cold backup system showed signs of failure as soon as it booted up. The simultaneous failure of three independent systems was an incredibly improbable event, yet it appeared that that was exactly what we were facing. And while we knew how to manage a single failure, a triple failure that left us with no functioning C&C computers and no reliable means to command the ISS systems constituted a full-blown emergency.

I was confident the team would find a solution but concerned about whether they would have enough time to do so before the shuttle had to undock. After walking down to the shuttle flight control room, I asked the

> ### Redundancy Management
>
> The manner in which the backup systems are applied is called redundancy management. Systems that must be ready to operate at any time typically have three sets of hardware so they can tolerate two failures. The first unit is called the primary and is operating. The backup unit is typically powered up and ready to begin functioning within a few minutes, which is what we call "hot standby." The third unit is connected and ready to be turned on, but it is powered off to conserve its lifetime until the primary fails and the backup becomes the primary. This is called the "cold backup."

flight director to reduce the shuttle's power consumption so we could extend their docked mission if needed. His team was already working on it. Their quick action would allow us to extend the mission by at least one day, if necessary, and probably two.

Then, I went downstairs to the MER, where the operations and engineering teams were already reviewing what had happened and generating recovery options. I sat in the back of the room, paying close attention as the team analyzed the system. They laid out the short-term impacts of the failure and described the more serious problems we would face when the shuttle undocked. The MER team thoroughly reviewed the system design, test history, and anything else that might have contributed to the problem but found nothing to explain the failures. We needed good answers—quickly!

Word of the problem spread rapidly through the community, and the control center began filling with people ready to help. Without a second thought, everyone who could contribute to the effort showed up. Some stayed only for a few hours, while others virtually lived there until the problem was solved.

These people knew everything there was to know about ISS and its systems, and they were completely focused on finding a solution to this supposedly impossible failure. To give us a little more capability while the problem was being addressed, they decided to activate a little-known feature that had been included in the design "just in case." In addition to the three redundant C&C computers, the ISS used a network of small computers, which had plenty of unused computing power that could be called upon in an emergency, as communication nodes. Buried in one of these nodes was a last-ditch commanding routine called Mighty Mouse, named after a 1960s cartoon mouse superhero whose tagline was "Here I come to save the day!" We would soon find out if it could live up to its namesake.

As night fell, the flight control team was ready to activate Mighty Mouse. The crew had already gone to bed after their long day, but they had left a light on in the *Destiny* lab to test whether Mighty Mouse could execute a command from the ground. We all held our breath as the command was sent to turn the light off.

The light went off. A muted cheer swelled in the control room. After ensuring that nothing unexpected had happened, Flight Director Mark Ferring directed another command to turn the light back on. The light came on. Mighty Mouse had not really saved the day, as it could not fully replace the C&C computer, but this modest success boosted our confidence.

After congratulating the flight controllers, I returned to the MER's quiet beehive of activity. Engineers and off-duty controllers worked together to find possible sources of the failure and ways around it. It was slow, painstaking work, but the team was focused and thorough. The air was electric, and I wanted to be a part of it—but my role required a clear head for the next problem, not just this one. Sometime after midnight, I reluctantly went home to get a few hours of sleep.

Returning to the MER before dawn, I got an update on the problem. For some unknown reason, two of the computers had lost access to their hard drives and all stored data. The third computer seemed to be functioning internally, but it had had a failure in a card required to communicate with the rest of the system. Although our problem was far from solved, at least we now understood its scope.

We also knew that we would have to extend the shuttle's stay by at least one day, so I began the formal process to do so. The shuttle was already conserving its resources, so my next task was to delay the launch of the Soyuz, which was planned to follow the shuttle's departure. Our rules required a twenty-four-hour gap between a shuttle departure and the arrival of a Soyuz to give everyone time to recover from the workload and sleep shifting that was often required.

At the morning MMT meeting, we reviewed the status of the problem and its implications for the mission. Although the Russians were aware of the failure, they did not understand its implications. I carefully explained what was going on and asked them to be prepared to delay the Soyuz launch if necessary.

To my amazement, Victor Blagov dismissed the request as unnecessary. I again walked him through the difficulties we had to overcome before we could undock, repeated that we had already extended the shuttle stay by one day, and reminded him that the flight rule required delaying the Soyuz launch as well. Again, Victor appeared to ignore my concerns and refused to discuss rescheduling the launch.

Muting the microphone, I asked my support team to set up a telecon with Victor for later that morning. At the appointed time, we gathered in my office at the control center. Joining me were the NASA increment lead, the flight director, an avionics expert, and an interpreter. In Russia, Solovyov had joined Blagov.

We walked through the situation once more. The NASA flight director explained the problem and how the shuttle was essential for recovery. The avionics lead clarified that we needed to build a new software load for the computer and required the shuttle's high-speed data communications to uplink it. These factors determined when the shuttle could depart, and by flight rule and policy, the Soyuz could not dock while the shuttle was still at ISS.

Solovyov spoke first, insisting that we did not need to keep the shuttle there and offering to let us use their communications system to upload the new software. He pointed out that they had done something similar many times on the Mir to rebuild the attitude control software after its many problems. This answer caught me by surprise. I was glad that he was willing to offer up their facilities for our use, showing that he was trying to be part of the solution. But his approach would not work for us. While the modest Russian computers ran relatively simple software that could be uploaded in sections, our computers ran much larger programs that could not. Even if we could partition the software into smaller loads, it would take weeks to modify, upload, and integrate it into the necessary files. The C&C was needed for attitude control the moment the shuttle departed.

After I finished my explanation, Blagov informed us that Solovyov had left the telecon without listening to our arguments. Everyone knew that Solovyov was brilliant and had mastered the Russian system inside and out, but we also knew that he was ignorant of the American segment and completely uninterested in learning about it. We were facing the definitive test of whether our two proud groups could avoid the kinds of mistakes that had been made in Phase 1. It was not looking good.

Struggling to control my anger, I told Victor that he needed to get his management to reconsider and that we could soon be facing an existential crisis. Sounding oddly distant, he said he would discuss it and then ended the call. I was stunned by their intransigence and could only hope that Victor would remember our conversation about the Mir collision in his office.

As the day wore on, I wondered if I was asking too much of Blagov. Although the treaty that had created ISS stated that he and his team were to follow my direction, he had grown up under the Soviet system and was under the direct command of Solovyov. Solovyov, in turn, was directly controlled by Yuri Semenov, the head of RSC Energia. Energia was a private company that actually ran the Russian Space Agency's program. They were the government agency's brain trust, but since the fall of the Soviet Union, the company's profitability had

> **Common Mode Failures**
> Reliability can be achieved in many ways, with one of the most common being the use of redundant systems so that one can take over when another fails. This is very effective except when there is a common flaw in the design that all the systems can exhibit. In the case of the C&C computers, two of the three failures were common, caused by the heads in their hard drives repeatedly crashing into the disks until they were welded in place. The third failure was unrelated. Following this incident, the hard drives were replaced with solid-state memory that eliminated this common failure mode.

also kept the Russian program alive despite the government's poverty. I suffered no illusions that Semenov would accept direction from NASA, let alone me.

Years later, I learned that Solovyov had thought I'd been lying to him. After leaving our telecon, he'd sent his engineers to look at the data coming down from the station. Our team had shown him that there was no data at all because the computers were not talking to the communications system. Our hard evidence had made no difference; Solovyov would not trust Americans.

This was exactly the crisis I had been trying to avoid. Serious technical problems were inevitable in such a challenging program, but we were confident we could solve them if everyone worked in good faith. Adding Russia's capabilities had increased our flexibility in dealing with problems, but without trust and cooperation, strength turned into weakness.

When NASA had been ordered to include the Russians in the program, we'd taken steps to protect against the possibility that they might not be willing or able to follow through. We had no way to prepare for the risk of them refusing to cooperate.

Following the telecon, my team and I reviewed the conversation and tried to understand why Solovyov was being so unreasonable. The answer became clear when we remembered that this Soyuz was going to carry Dennis Tito. His fee had paid most, if not all, of the costs of the flight. From their perspective, taking care of a paying customer was more important than dealing with our system failures.

With that insight, I went to see Holloway. After telling him what had happened, I recommended that we ask for help from as high up as we could get. Just a few months earlier, that would have included Vice President Gore, who'd had a personal interest in the ISS and an excellent relationship with his Russian counterpart. But by the spring of 2001, the administration had changed, and the Bush White House had neither the interest nor the relationship to be helpful.

Tommy had his secretary call Dan Goldin, the NASA administrator. Dan was aware of the emergency and took our call immediately. We explained the problem and asked for help.

"I understand," he said. "I will call Koptev immediately and ask him to go to Baikonur to make a direct appeal to Semenov." Yuri Koptev was Goldin's counterpart in the Russian Space Agency.

We thanked him and hung up.

"I have no doubt Goldin will call," Tommy said quietly, "but I doubt it will have much effect. You had better find a way to live with the Soyuz arriving on schedule."

The next day, Holloway came to see me in the control center. The look on his face said he had bad news. We went into my office and closed the door.

"Goldin called to say that Koptev had no luck. Semenov said something like, 'There is not one ruble of government money in this launch, and it goes when I say it goes!' Then he gave Koptev back a watch that Koptev had given him as a gift." Returning a gift was a grave Russian insult.

I did not know what to say, and after a moment, Tommy went on. "How long can the Soyuz hold out before docking?"

"I don't know exactly, but it could be a few days if they don't try to rendezvous too soon. On the other hand, if they go ahead with the rendezvous and have to hold position, they may use up their maneuvering fuel pretty quickly and have to go home without docking." I reminded him that the Soyuz currently docked at the station—which was the crew's means of leaving in an emergency—would reach its authorized lifetime in orbit in a couple of weeks. According to the plan, it would be swapped for the one coming up. If the new Soyuz did not have enough fuel to dock, the station crew would have to bring the old one back. They could not stay aboard the station without a means of emergency return. "The station is not designed to be run without a crew on board," I concluded.

Tommy nodded grimly before he spoke.

"Dittemore has already said that as long as he is running the Space Shuttle program, he won't let the Soyuz dock while the shuttle is there. You need to work something out. If they launch and can't dock, it could be the end of the program."

I was certain that if presented with the facts, Goldin would order Dittemore to let the Soyuz dock, but that was an ugly option that would not solve the problem. The station would struggle to control itself with this unexpected configuration, rapidly consuming fuel needed for other purposes.

When Tommy left, I felt completely helpless. After a few minutes, I called in our senior interpreter to call Victor Blagov at home.

"Victor, I am sorry to bother you this late, but I desperately need your help. We have an absolute crisis here—we need to delay the Soyuz launch."

"Jim," he said, "I understand what you are telling me, but what you ask is impossible. Semenov himself is overseeing the launch to ensure it goes as planned. He will not change his mind. But it is not that bad. We can dock while the shuttle is there, and they can stay as long as you want."

"You don't understand. The Space Shuttle program has people whose heads are as hard as any rock in Russia. Their leader has already said they will not allow the Soyuz to dock while the shuttle is there, and they can hold the station in attitudes that you cannot use. And even if I can change their minds, the station can't maintain attitude control with the two vehicles attached. It will burn up all the fuel, and then we will truly be lost. We need time to recover the computer so that the station can control itself."

Victor said nothing, so I went on. "Victor, if we can't stop the launch, perhaps we can delay the docking. We've already agreed to let the shuttle extend its stay by one day without impacting your plans, but if that's not enough, I may need you to help me by delaying the rendezvous. This is absolutely critical. You know that if the Soyuz performs the rendezvous but cannot dock, you will burn up its fuel and have to bring it home. That would disappoint your customer and really get Semenov angry. Worse yet, when the current Soyuz has to leave, the station will be unmanned. The American segment needs a crew to survive, so going unmanned could kill the program."

Victor waited a long moment before speaking.

"Jim, you have always told me the truth, and I believe you when you tell me this. I will see what I can do. But Soyuz will launch tomorrow as planned. Call me Sunday morning before I go to work. If you still need one more day, I will put an error into the rendezvous command that is sent to the Soyuz. The burn to begin the rendezvous will abort, and you will have the extra day you need."

This was a genuine breakthrough. Not only was Victor offering relief for my current problem, but he was also putting our partnership ahead of his own interests. As a Russian in his sixties, he had every right to be thinking about retirement. Instead, he was offering to put his future—and that of his family—at risk to save our partnership. Nothing could have spoken more powerfully about the changes rippling through our world.

"Thank you, Victor! I know that this is difficult for you, and we will do all we can to solve the problem without your help, but you may be our last hope."

The next day, the Soyuz launched as scheduled. Our engineers were making progress, and we were becoming more hopeful, but there were still issues to be resolved. Even if they succeeded and we could let the Soyuz dock normally, this would violate the flight rule for time off between events. I asked MOD and the crew office to make an exception, and after talking to the crew on orbit, they agreed.

We had already identified the minimum level of computer capabilities needed before we could allow the shuttle to undock. Late in the afternoon, the flight director and MER manager came to see me.

"We have met our minimum configuration. We don't like it because it is absolutely bare-bones minimum, but we are okay to undock the shuttle tomorrow as planned."

I asked them to review the plan and the remaining risks for me. When they were done, I thanked them and let them go. I asked the interpreter to be ready for a three-way telecon with Victor Blagov at eleven o'clock in the evening Houston time. I called from my house.

"Victor, it is Jim. We have fixed the computer well enough to undock the shuttle tomorrow. The Soyuz can dock!"

"Thank you, Jim!" he said with obvious relief. "We will both have jobs tomorrow!"

32. SETTLING DOWN

The C&C computer worked fine, and we once again had control of the ISS systems. The shuttle departed as scheduled, and four hours later, the Soyuz docked. Dennis Tito had his tourist stay on the ISS, apparently oblivious to the crisis he had precipitated.

The breakdown of trust during the crisis revealed a weakness in our relationship, but I forced myself to embrace the positive. Our political masters had made it clear that their support depended on our cooperation with the Russians. This existential crisis had tested both sides to their limits, but the relationship had held—if only just. Victor Blagov's trust in me and his remarkable willingness to put his career at risk showed that he, at least, saw that our futures were tied together and that cooperation was the only way to protect his beloved Russian program. This was a huge step forward.

Tommy Holloway could not see it. He and others in his career cohort had matured during the glory days of NASA's unquestioned political support. Those were the days when they could make the nation's space program what they wanted it to be. In retrospect, I think he expected the cooperation with the Russians to fail—perhaps even wanted it to—so we could revert to the more comfortable ways of the past. He had not endured the near-catastrophe of Freedom or the wrenching trauma of 1993 in Crystal City, when the entire space station initiative, the Space Shuttle program, and perhaps NASA's great adventure in human spaceflight had dangled by a thread. But NASA's glory days of unlimited budgets were long gone, and I could only stand astonished as he lectured me about the Russians doing whatever they wanted and called me foolish for having trusted them.

After the excitement of 6A, we were grateful to have a period of relative calm before the next assembly flight. While we kept busy activating and checking out our new systems, the relatively slow pace gave us a chance to carefully assess how well things were performing. Most of what we found was good news, with the majority of systems working almost perfectly.

Some of our most difficult challenges came from completely unexpected sources—such as a simple test to verify that the wastewater dump system would function as designed. While NASA would eventually recycle most of the water on the station, until those systems became operational, we would vent wastewater overboard. NASA had used water dumps on most of the human spacecraft we had built, including the shuttle, so we considered the test routine.

But Russian specialists had reservations. They had never liked the water dumps that had occurred when the shuttle had been docked to Mir and had only tolerated them because they had been rare. They were much less willing to accept them as a routine event on ISS.

I also discovered that Jay Greene had encouraged the NASA team to ignore the Russians' concerns. He expected me to overrule them when the issue finally came to the MMT. I would not do so unless it was an emergency, as doing so would compound the damage Jay had already done. I had to correct that mindset before a serious error resulted.

> ### Managing Waste Water
> Short-duration missions like Apollo and shuttle flights could afford to dispose of wastewater by dumping it overboard. The amount of water used was small enough to make it easier than trying to recycle it. But long-duration missions, like going to Mars, will require that the water be recycled. Although ISS was developing systems to recycle the water, those would come years later.

There was also a real possibility that the Russians could be right in their concerns about contamination and ice debris. Although I had plenty of experience with argumentative Russians, they often had a valid point. My duty included protecting their interests as our partner and learning as much as we could from their experience. I asked the MER to plan a test that would protect the Russian segment from contamination by directing the water flow opposite our direction of movement. The Russians were relieved to have their concerns respected.

We positioned the SSRMS so that the cameras on its end effector would capture the spray pattern coming from the nozzle. We all expected it to resemble a fine mist from a garden hose. Sitting in the MER conference room, we chatted easily as we waited for the flow of water to begin. As soon as the first drops appeared, the room was filled with shocked surprise. Instead of a fine spray flowing in a tight pattern, we saw a broken, swirling stream of large droplets spewing forth. Droplets this large could flash freeze into iceballs that might harm the station if encountered later. We were stunned and embarrassed; this was not what had been expected, and nobody had an explanation for why.

At the end of the test, the NASA team went off to review the design. The Russians watched the video but had no idea what was expected, so they had no immediate reaction. I was glad for their silence; making a fuss to embarrass the Americans would only have made it worse. For me, the test reaffirmed my long-standing skepticism about our ability to perfectly model things we had never tried before.

A few days later, the NASA engineers had a theory about what we had seen. They did not explain why their prediction had been so flawed, but the incident did make them more willing to listen to the Russians' concerns. I hoped this was one more tiny step toward the trust and cooperation we needed.

The slower pace let us look at many things more thoughtfully, often uncovering new twists in our knowledge. Another baffling phenomenon we observed was the way Russian EVAs perturbed the

ISS attitude control as if something was pushing on it. No such perturbations appeared when the NASA suits were used. Many factors
were considered, such as an air leak from the Russian airlock, but no
clues to the source could be found.

Many months later, Jack Bacon, a physicist, found an answer in
a detail nobody had considered significant. Jack is one of those tenacious characters for which NASA is famous, and he refused to let go
of the issue until he had an answer. Because it was a low priority, it
took him a long time to chase it down, but he did.

A space suit must get rid of heat from both the human body
and the equipment it contains. Both Russian and American suits
accomplish this by sublimating water. In a device called a sublimator, water flash freezes to ice in the vacuum of space, then subli-
mates into vapor. Although the rate at which the water is released
is tiny, it theoretically produces thrust. For this reason, the American suit deflects the vapor out of both sides of the sublimator to
cancel any thrust it might create. Because the effect is so small, the
Russian design vents all the vapor in the same direction. Jack demonstrated that this tiny thrust had produced the perturbations we
had seen. When he presented his analysis to me, I could not believe
that a few water molecules could move a three-hundred-thousand-
pound station.

"What do you think the velocity of the sublimating water molecules is?" he asked.

"I have no idea. Maybe a few hundred feet per second?"

"Three thousand feet per second," he said, letting the number sink
in. "If the crew spends a couple of hours working on the same side of
the station, you have something like ten pounds of water being ejected
in the same direction. Don't you think you would feel the effect of a
ten-pound block of ice being fired at three thousand feet per second?
The same thing is happening during an EVA—same net thrust, just
spread out over hours."

I loved it when we learned new things.

We also had to deal with a new category of human challenges. Within a few weeks, two astronauts who were good friends of our crew on ISS died on Earth, and we had to carefully share the news. Dave Walker had lost his fight with cancer, and Patricia Robertson had died in an airplane accident.

We were prepared for this in part because we had seen it go wrong on Mir. The Russians had made a point of asking their crews if they wanted to be notified of the death of a family member while on orbit, and most had said no. During Norm Thagard's stay, one of his Russian crewmates had become deeply upset when he'd learned of his mother's death during an American media event. We needed a way to share difficult news before it leaked, so the crew office and medical operations worked it out. It was still a difficult conversation, but the approach worked.

As the weeks went on, the team worked through the list of hardware and software anomalies, fixing what needed repair, and our concerns gradually eased. We finally had redundancy in all major functions and enough performance margin to get through almost any problem we might encounter. Our roller coaster ride seemed to be smoothing out. The ISS needed only one more major item before it could take care of itself: an airlock.

We had been relying on the shuttle's airlock to work on the American segment, so the crew could only perform EVA maintenance during shuttle flights. The Russians had limited EVA capability through their modules, but their suits and equipment were not designed to work on the US segment. For the research and maintenance work that lay ahead, we needed the fully capable US airlock.

Once it was delivered and activated, we would officially reach the end of Phase 2, also known as the Assembly Critical Phase, which had been my focus since Freedom.

33. HITTING THE RAPIDS

John Lennon said, "Life is what happens to you while you're busy making other plans." With the end of Phase 2 in sight, we were well into planning Phase 3, which would convert our stubby runt of a station into a productive research facility with a fully outfitted *Destiny* lab, dual laboratories from our European and Japanese partners, and a window-walled cupola with a breathtaking view of Earth. Even though I knew I was living what was likely to be the most exhilarating and rewarding time of my life, my days passed in a blur.

The team was maturing into what the program would need for the long haul. Many new faces were showing up, eager to help, including a few from the Space Shuttle program. There were signs that the Space Shuttle program team was slowly learning to accept ISS as a program with equal stature and validity, and some of its people could even see that we were creating a lasting partnership that could do great things far into the future.

Many attitudes seemed to have changed because of the crisis during 6A. When the problem had first emerged, it had seemed like another of those great moments when NASA engineers would step up to solve a technical emergency. Although the team had responded as it always did, the problem had needed a response that was more than technical. The critical element had been my connection with Victor Blagov—the payoff from years spent earning his trust and respect.

317

After it was over, several flight directors told me they'd found the experience intimidating. While they were happy to manage the vehicle, they were even happier to leave the human issues to me.

After hearing one of the proudest flight directors admit this, my colleague from Phase 1, Missy Gard, told me quietly, "You know, that was as close as a flight director will ever come to giving management credit for a successful mission."

Some attitudes on the Russian side also began to shift. Blagov was clearly embarrassed that Russian leadership had shown such indifference to our problems. Although he and I had never been close, he showed a new respect for my efforts to hold the partnership together. My interpreter friends said he could see that I treated all parties as equals, sharing credit and blame as warranted.

Still, I sensed simmering organizational tensions on the American side standing in the way of progress. I suspected many of them traced back to when Freedom had been canceled. That ugly time had been made much worse by the clumsy and insensitive way Goldin had punished the innocents for the errors of their leaders. The damage would take years to repair. And, as the only senior manager to transition directly from Freedom to ISS, I felt a survivor's guilt.

From time to time, I would come upon former colleagues who had been driven from professional positions on Freedom to low-paying jobs elsewhere. One particularly difficult encounter grew out of a chance meeting in the Baybrook Mall, about five miles from JSC, on Christmas Eve. I ran into a former McDonnell Douglas colleague who had once been a respected middle manager and had been reduced to working part-time at RadioShack.

Other Freedom managers had come back into the ISS fold after some time in purgatory. Most did their best to help us, but a few found ways to make their hard feelings known. One who was at headquarters delighted in questioning all my decisions. His criticisms did no damage, but life would have been easier had I received a little respect.

While I felt deep sympathy for those who had been hurt by that brutal transition, I had none for those who threatened ISS by clinging to their personal power base. The most common form of this came from middle managers in the Space Shuttle program, who made our lives difficult every chance they got. Where most people would have been delighted to have their responsible and challenging jobs, they used every opportunity to lord their power over others.

Inevitably, the conflicts within the ISS team grew as more people got involved. The early flights had been conducted by a relatively small team that had done amazing things fueled by adrenaline. Most had been painfully aware of just how fragile our new program was and had worked tirelessly to protect our partnership. The newer people neither understood nor respected that underlying fragility. Keeping them in line and sometimes repairing the damage they created took a disappointing amount of time and energy.

When this seemingly endless effort got me down, I would spend time in the trenches with the folks who put their hearts into their work. Their selfless commitment to our historic effort inspired me and recharged my emotional batteries.

As we worked past the adrenaline-fueled stage, I started to see some of the team showing signs of fatigue. Where I was used to respectful comments like "He didn't get to it, so I took care of it," I was beginning to hear comments that showed unhappiness with colleagues. I was particularly shocked when one flight director leveled an attack on her colleague that ended with "I wouldn't piss on him if his guts were on fire!"

This erosion of collegiality led me to take a closer look at what was going on across the organization, including my own office. In some cases, I could now see people directing their understandable frustration toward their colleagues. I immediately felt guilty that I had not seen this earlier.

My wife reminded me that I had been working at the limit of my ability and assured me that my only "guilt" lay in treating people like adults. And while most people vindicated that trust in spades, a few did not.

Even the outstanding team in my own office was being stretched to the limit. Susan Creasy was bearing most of the burden of the mission integration work for the upcoming missions while being my close counsel on all issues. She not only had the mountain of planning work well in hand but would occasionally step in to run the MMT when I could not. She was doing a fantastic job, but the strain was beginning to show.

Chirold Epp also did yeoman's work and led the operations management effort. He oversaw the huge budget and worked tirelessly to keep my monthly review down to a day and a half. There were dozens of major issues that kept him constantly challenged.

Leading and integrating the operations program was an enormous task that required a strong management team, including backups for all critical personnel. The reduced management team that Holloway had forced on me—particularly his leaving me without a deputy—did not provide the depth needed to deal with the years of stress that lay ahead.

My workload was taking a toll on my health. I could foresee a time when I would have to step back and take on more of an oversight and leadership role. The time had come to start grooming replacements for myself, Susan, and Chirold. This would have been much easier if we all had deputies.

I knew Holloway would resist such a request, so I set out to recruit someone who could overcome his objections. There were some candidates within my shop, but none were an obvious fit, and all were busy with their own work. I decided to reach out to some flight directors to see if any were interested in moving up to management.

Bob Castle—my mentor and guide into the world of flight operations when I'd been detailed to the office during Freedom—had an even temper and lots of experience with the Russians. When I asked him about coming over, he said he was still having too much fun at the flight director's console to leave.

I also asked Wayne Hale, one of the most senior flight directors, whose technical expertise stood out even among that august group. I knew he would quickly gain the respect of the Russians.

"I am definitely looking for an opportunity to move into management," Wayne told me, "but I want an SES (senior executive service) position. You're not an SES, and your deputy won't be either, so there's no obvious path to a promotion."

He was right, of course. When Shep had selected his management team at the beginning of the program, I'd been promised an SES within a year or two because it was the proper grade for such a leadership role. But a decade later, that broken promise had put my ability to orchestrate a smooth succession—to say nothing of my health—on the line. It was time to take action.

Rather than fighting with Holloway, I chose to go talk to the head of human resources. Greg Hayes had been a division chief when Dennis Webb had offered me the job at JSC. I knew he hadn't liked me even then, but he had followed orders to bring me on board.

When I explained my difficulty in recruiting a successor, he responded in a curious way.

"There is no doubt that your job justifies an SES," he said forcefully, "if the right person is in the job."

We talked for about twenty minutes, during which he emphatically repeated that phrase—"right person"—several more times. I did not understand what he was implying. Having been asked to take the job over all other candidates and having done it well for several years, it had never occurred to me that he was implying I was unqualified. His indifference to my plight was particularly disconcerting because virtually everyone else I knew was sympathetic, supportive, and eager to help me and the program. Only later did a friend who worked for Greg tell me that he had long ago decided to do everything in his power to block my career. Although I was hurt that he had so little respect for me, I was even more upset that his contempt would hinder my ability to do the very best job I could for the program—both now and in shaping the next generation of leadership.

After that dead end, I went to see Holloway.

"You don't need a deputy," he said. "You need to back off. You are one of the hardest-working people I have ever seen, and you're going to burn yourself out unless you ease off."

"You don't have to talk to me about burnout," I snapped, "but there is no way I can cut back on my workload. Every day brings new problems, starting with holding the team together. You know there's nobody else the Russians are going to listen to right now!"

He had to know that if I didn't do what needed to be done, it wouldn't get done. I couldn't accept that he would think that was okay.

Tommy only repeated that I should "stop working so hard," but he didn't say a word about a promotion. So, I did.

"Look," I said, staring hard at my boss. "You know full well how critical my role is to this program. I'm the glue holding this partnership together. I practically live in the control center, spend more time in the center director's office than you do, and am directly responsible for the safety of NASA's most precious resource. I bear more responsibility and exercise more authority than almost any SES in the agency, yet I have been denied the recognition of that title. So, it's time to balance the books. If I'm incompetent, you owe it to the program to fire me, but if I'm doing a good job, you owe it to me to promote me!"

Holloway was startled by the anger and force of my argument, and he looked away from my stare. Even as he grumbled about my "working too hard," it was clear I had made my point. He said he would put me in for the promotion, warning me that it would take some time.

Despite his reluctance, I used this grudging sign of support to rebuild a little of my morale. The technical challenges of my job were difficult, but the human ones were dispiriting.

And the human challenges never ended. Many of them popped up when the daily plan was published at the beginning of the crew's workday. Since the crew worked on Moscow time, I got lots of phone calls in the middle of the night. There were still many disputes among the middle managers who felt their authority was being tested. Rather than trying to work things out with their Russian counterparts, some flight directors would demand that I "make the Russians do what we want." Just as often, I would get messages from Victor asking me to "make the Americans come to their senses." The operational

> ## The Crew's Workday
> Early in the program, the crew's workday had been set to coincide with the workday in Moscow. The Russian space program paid its people so poorly that the flight controllers could not live off of their salaries. They needed second jobs just to keep their families fed. If we made them work on Houston time, many of the controllers would have to give up their space jobs. The loss of expertise would have been devastating. The workday was later changed to coincide with London time, roughly halfway between the two.

pace meant I had to be decisive, but the fragile partnership meant that I had to stay within the bounds of what people would accept. Painful experience taught me that the only thing to do was listen to both sides and decide on the technical merits. It turned out that each side was right about as often as they were wrong. This meant that I made nobody consistently happy.

These constant squabbles kept me on edge, and knowing Holloway would not back me up left me feeling utterly alone. Randy Stone was very supportive, but he had no authority to enforce my decisions.

My patience slowly eroded. Many times, I wanted to yell, "Just grow up, people!" though I never did. Colleagues warned that by not putting people in their places, I was training them to whine while bearing the whole burden of finding common ground, but I could not see an alternative. Our future required mutual trust and respect above all, and that meant listening to everyone.

When Jay Greene, who had been responsible for demoting my position, finally left the program, Bill Gerstenmaier came down from the Space Shuttle program to replace him as Holloway's deputy. We had often been in conflict while he'd been my direct ISS counterpart at the Space Shuttle program, but I allowed myself to hope that his Phase 1 experience and Space Shuttle program connections would lead to better relationships and processes when he joined our team. That did not happen. He seemed to favor the Space Shuttle program approach even when it hurt us. Worst of all, he tried to run ISS the same way he had the shuttle program. His arrival did nothing to improve my relationship with either him or Holloway.

Because he had held the equivalent role to mine in the Space Shuttle program—operations manager—he should have realized that I had crafted a sophisticated program to get us through these dangerous times. But instead of respecting my work, he picked up where he'd left off in the Space Shuttle program and tried to run ISS as if I did not exist. He never approached me to understand my roles and responsibilities, and we were soon in conflict.

I asked him into my office.

"Bill, I have a real problem with what you are doing. You seem to forget that I am in charge of ISS operations, not you. You are my boss and supposed to support me, not mess up the plan. If that is not how you see it or if you don't have confidence in what I am doing, you need to speak up. And if you want to do my job for me, then you can have it. Lord knows I am sick of working all day, every day, while taking fire from every direction. So, either take my job or get out of my way!"

Bill was shocked by my outburst, but he immediately saw my point. He quickly apologized and promised not to interfere anymore. While I am sure he meant it, it was a promise that he had trouble keeping. We would have more discussions.

The new stress from Bill was the last straw. I had to admit that I could not keep up the workload I was bearing. I went to Holloway.

"I think it is time that I give the mission integration part of my job to a deputy and focus on being the MMT chair and head of operations. I'd like Susan Creasy to be promoted to head the OC office so that I can come back to the front office to do the ops integration job as it was originally set up. It would also make it easier to justify the deputy role."

Tommy was quiet for a moment. "I agree that you are trying to do too much," he said, "but I am not going to create the deputy job. You can come over here, but the title will still be manager of operations."

I had pretty much expected this from Holloway. He had a reputation for berating others while asking them to do more. His management style was costing us dearly, but there was nothing I could do about it at the moment.

I accepted the move, even as I could see the distant horizon of my time with ISS.

34. COMPLETING PHASE 2

Susan was sad but not surprised when I told her about my move. She knew better than anyone that I could not continue much longer. We had only worked together for a few years, but our teamwork had been spectacular. She had done everything she could to support me—giving me her best advice in private and her total support in public. I could not have asked for a better deputy or friend.

That partnership would certainly continue in our new relationship, but she would finally have the chance to get credit for her leadership. It was a role she had long deserved, and I was glad she would receive some of the recognition she was due.

Chirold was also growing tired of his role in the program and was looking for other opportunities. He would soon go back to the Engineering Directorate to work on exploration missions, one of his favorite tasks.

Settling into my new office in Building 1, I had a strange sense of being in a temporary job. The slightly less frantic pace gave me time to appreciate how our team was finally getting into the groove. Surprises were becoming fewer, and problems were beginning to fall into patterns that we knew how to handle. The stress level decreased as things became more routine. It was good.

Each successful flight also made the station better able to handle problems that might arise. Flight 7A delivered and installed the airlock as smoothly as any flight I had experienced. With it and our growing supply of spare parts stored on the station, we were finally in a position to recover quickly from many failures. I saw the fulfillment of the goal I had set so many years earlier as the maintenance manager for Freedom.

The station was performing beautifully, and our audacious plans were becoming a reality. Media interest was high, and the coverage was positive. People who had devoted a decade or more of their careers to the space station finally got to see their hardware fly.

The progress sustained the team's morale. Some people whose hardware had launched were rejuvenated by seeing their efforts rewarded and eagerly jumped back in for another round. Others, worn out from years of frustration and uncertainty, enjoyed the moment and then moved on to new jobs.

My own work left me feeling profoundly conflicted. I was thrilled to see the ISS coming together so well and immensely proud of the men and women who were making it happen. Yet, the endless arguments across the organizations wore on me. The most difficult part remained the lack of support from my boss. A good boss would have shared our successes and built a sense of teamwork, but Holloway continued to snipe at me. He never offered the support I should have had, instead creating a growing emptiness where I needed a joyful sense of accomplishment. I felt abandoned and alone.

Although we did not dwell on it, I knew my wife was beginning to wonder how much longer this would go on. She and I had always been intellectually and emotionally close, but the demands of the last few years had tested that closeness severely. My work often took everything I could give, leaving my family to see the grouchy irritation I could not show at work. I sometimes wondered how, as I grew distant and emotionally drained, my family could continue to love me during our time together.

While I knew this could not continue indefinitely, no event on the schedule suggested a good time to leave, and no individual stood out as a natural successor. The work remained demanding, but it was also invigorating. I lived in a confused muddle of staggering costs blended with profound rewards.

It seemed most likely that my health would force my decision. I was living with a growing list of stress-related maladies that followed me home and left me utterly drained at the end of the day.

My move back to Building 1 gave me just a little more time for myself. I could more often get home in time for a late dinner and a walk with my wife. On weekends, when the shuttle was not docked, I was often home by early afternoon. The needs of the station were more routine, and the team was better prepared. I still took no holidays or vacations. My family did their best to cheerfully put up with it, even as I kept promising that it would end soon—but they knew this was more a hope than a plan.

Doris had the roughest duty, patiently tolerating my bad moods and looking for ways to support me. She found books on stress management, soothed me with Beethoven and Mozart, and never let me doubt her love. She helped me discover new techniques to manage the stress and maintain my sanity.

When that sanity was being tested by the bad behavior of others, Doris taught me not to attach motivations to behaviors. She pointed out that although I could often predict those bad behaviors, I could never really know what was motivating them. This helped me treat them as if they were hardware anomalies—objects behaving according to their own inevitable laws rather than people acting with malicious intent—which helped a lot.

Doris would often observe what a remarkable set of circumstances had brought me to this point and how, in retrospect, it almost seemed inevitable. I agreed that it would have represent the greatest of good fortune if I only had a boss who supported me.

One day, after a particularly draining MMT meeting, Tom Marshburn, one of my flight surgeon friends, came over looking concerned.

"You doing okay, Jim? You look exhausted." Tom was always thoughtful, and I appreciated his inquiry.

"Actually, I'm working on beating Sonny Carter's record." Sonny was an astronaut famous for getting through medical school on two hours of sleep a night. "Soon, I'll be getting by on no sleep at all!"

Tom scowled. "You need to take better care of yourself. If you take my advice, you'll take

---

Little Jimmy

One of my wife's techniques for helping me keep my situation in perspective was to imagine somehow visiting myself at age fourteen, telling my young self about my current job. "Just think how little Jimmy would have felt if he knew where he would end up!" she would say. That fanciful thought helped me see past my problems and appreciate the many rewards of my job.

---

some time off and unwind a bit. If you want some suggestions on how to manage the stress, I'd be happy to get you to someone who can help. We need you to be healthy—almost as much as your family does!"

I promised him—and myself—that I would work on taking more time off. As it turned out, others wanted that for their own reasons.

One morning, when I entered the MER, the manager pulled me aside.

"Can you start coming in at seven in the morning instead of six? We have to brief our management by phone before you show up, and they are tired of getting up so early."

The request surprised me. Mission events had me in the control center around the clock, and there was no way to filter my access to the data. The early hour was driven by the fact that the crew day was already well underway, and I needed to get up to speed quickly. Moreover, the idea that I should reduce my own preparation for someone else's comfort was not very professional. Still, I was tired, and maybe this was an opportunity to ease off a bit.

"Actually," I said, "I was thinking of going back to a five o'clock showtime." That had been our routine in the first weeks after the crew had arrived.

His face blanched, so I laughed before he had a stroke.

"How about 6:30? I can live with that as long as we don't have any serious problems and you keep your act together."

"I'll pass it on. Thanks."

Looking back, our success makes it seem as though I worried for nothing, but that is the nature of planning. If you do it well, things work so easily that it is hard to imagine anything going wrong. Poor plans have the opposite effect—making the initial effort easier but success almost impossible. My deep sense of personal responsibility meant I was driven to be ready for whatever might come. That constant focus on being prepared for the next problem made me hypersensitive to the issues. Others thought I was not giving myself credit for what was working. Either way, I could not turn away from the challenges that lay ahead.

One incident during this period gave me reason to hope that I might finally have found a way past Tommy's criticism and indifference. It came when the Russians wanted to do an EVA on the *Zvezda* module. The EVA crewmembers would be working close to the module's maneuvering thrusters, which had toxic combustion residue around them. The NASA safety and medical communities were concerned that the crew would bring it inside on their suits. The Russians were adamant that the EVA needed to take place, and the NASA safety and medical offices were equally adamant that it was too dangerous.

After days of working on an approach acceptable to both parties, I assembled the MMT to review and concur on the final plan. It involved procedures to keep the crew away from the contamination and ways to clean up if contamination did occur. Holloway had heard about the EVA and the complaints and came to see how it went. I am sure he expected me to be forced to capitulate to the Russians.

While the flight director gave a quick rundown of the planned EVA, I got up and went around the table, talking quietly to the reviewers, who would soon have their chance to ask questions and express their views. I made sure that each knew what was coming and insisted on their frank assessment of the plan. I also told them to be sure they understood every syllable of what we were proposing. Anyone who objected without having facts to back their position would undermine their influence on the decision.

Returning to my seat, I guided the discussion around the table to give every organization a chance to express their concerns without any pressure. In the end, each found the final plan acceptable. Victor Blagov's relief and appreciation were clear, even in his untranslated voice, as he acknowledged the clearance to proceed.

When the meeting broke up, folks came over to give feedback, update me on issues, and ask for input on other decisions still ahead of us. Holloway stayed seated and chatted with folks. As the crowd finally dispersed, he came over to me.

Speaking softly, as was his style when he was being serious, he said, "That was a thing of beauty." Then, after looking hard at me for a moment, he turned and left.

He had just given me the best compliment he would ever offer. Not one to flatter people—especially me—he had said what he'd had to say and left. It would have been even better if he had appreciated that this was my normal way of doing business, not an aberration. Unfortunately, he had not, and this did not herald a new approach from him. Instead, we continued to fight about how to manage the Russians.

35. BEGINNING PHASE 3

Our successful completion of Phase 2 was a tremendous accomplishment, even if it passed almost unnoticed. The steadily improving ability of the station to take care of itself relieved the sense of urgency that had been hanging over my work, but I remained careful not to become complacent. We had a lot of challenging work ahead of us.

Bill Gerstenmaier was trying to do better, and I appreciated it, but his old habits were difficult to break. We were frequently at odds. The unending problems with my management turned my thoughts more and more often to moving on to something else, though I had a hard time imagining a job that could compare to what I was doing. I was effectively leading human space operations for planet Earth, and that put me in the middle of every operationally significant discussion at JSC. Yet the lack of support from my bosses meant I was rapidly using up my physical and emotional reserves. It was time to think seriously about what should come next.

The first major challenge was finding a replacement whom the Russians would respect. There were plenty of people technically qualified to do the work but few with the needed mix of experience and credibility. Frank Culbertson, my old Phase 1 boss, would be a great choice after his mission and some time to recover, but that was more than a year in the future. Gerstenmaier could fill the bill, but he was too invested in doing things NASA's way. There were others who could take the reins, including Susan Creasy, but it was not obvious who was best.

I also wanted to see Frank safely through his mission. He was about to launch as the third commander of ISS, and even though our relationship had become strained after the end of Phase 1, I wanted to give him the smoothest mission possible.

Frank and his Russian crewmates launched on flight 7A.1, another of the many logistics missions that got added to the original assembly sequence. This was the third mission with an MPLM, and it went beautifully. As I watched science racks being transferred, I got a note that Tommy wanted to see me.

"I saw the pictures of the new equipment in the lab," he said. "When are we going to see the research program get started?"

"We have been doing some all along, mostly human research on the crews and photography from the science window because they don't require much hardware or crew time. This flight delivered some new hardware but not a lot of logistics, so our options are still limited. Besides, we still have a ton of work getting systems activated and working off the anomalies. Why do you ask?"

"The natives are getting restless," he said. "The science community says they supported the program as a research platform, and some of them are getting impatient. You have officially begun Phase 3, and they want to start their work."

I nodded. "I was part of all that and want to get them their time as soon as we can. We have research time built into every schedule, but it is a matter of balancing priorities. When things go well, we put in more research time, but when things go wrong, it has to be the first thing to go."

"What is the average now?" he asked.

"I would guess about ten hours a week. It varies widely based on the workload."

Tommy stiffened ever so slightly.

"I want you to find a way to give the researchers twenty hours of crew time each week," he said. "I know that there will be weeks when it will be hard, but we have to make a commitment."

"I'll do my best, but I can't guarantee anything," I said.

"That is not good enough. We need to do this. If you can't meet the twenty-hour minimum, you will have to explain why in front of the space station control board."

I didn't like this development, but it made sense. We desperately needed to keep the science community on our side, and all they were asking for was what had been promised. The fact that the promise had been made long before we understood the challenge was not enough.

I discussed the order with Missy Gard, now the increment manager for Frank's stay on the ISS.

"How hard is this going to be?" I asked.

"We can do it," she said quickly.

"You sure? We have a lot going on," I said.

"As long as we don't have any major failures or Russian fights, we will be okay. And Frank wants to do it. He will make time."

As usual, Missy was right. As the increment manager, she had both the big picture and the details that made it work. Missy had been a close confidant to both Frank and me since she'd joined the Phase 1 office years before, and I knew she was strongly motivated to make sure he had a successful mission.

I was amused that Holloway was expecting a change with the completion of Phase 2. He was otherwise indifferent to most of the programmatic agreements we had made, which probably meant that he was getting an earful from the researchers. I hoped that this meant he was beginning to understand the program.

In truth, the pressure to accommodate research was a relief from the tedium of routine operations. One of my personal concerns was that the combination of pressure and tedium would begin to affect my judgment. As a student of aerospace history, I knew that more accidents resulted from a gradual erosion of judgment than from a dramatic failure of knowledge. I had long ago developed a routine to assess the degree to which my decisions relied on gut feelings rather than objective information. As long as I could trace my decisions to hard data and firm logic, I knew my judgments were sound. I still felt confident that I was in total control of my faculties, but I was worried that my performance would slip without my seeing it.

One day, after Frank's crew was safely aboard, I got a note from Dom Gorie, commander of the upcoming UF-1 flight, wanting to talk about the amount of stowage in the shuttle mid-deck. Although I talked to every ISS crew, those days I did not usually see the shuttle crews except in the preflight and postflight meetings because we were all busy and had to focus on the things closest to our responsibilities.

Stowage was the Space Shuttle program's responsibility, and while I wanted them to accommodate all that we safely could, I did not normally get involved in the issue. To prepare for the meeting, I called MOD to see if they could shed any light on what was bothering Dom.

"Why don't you come over and see for yourself? It just so happens that we have a crew compartment trainer set up in the launch configuration for Dom's mission."

The crew compartment trainer was one of the full-scale mockups used for engineering studies, crew training, and public affairs events. This would let me see the same stowage configuration that Dom would have when he flew, explained by one of the instructors who taught his crew.

I clambered into the mid-deck, the compartment between the cockpit and the cargo bay that held most of the stowage and where the mission specialists sat for launch and entry. It was easy to see the issue. Even without the crew seats installed, the compartment was small. With the seats and seven crew members strapped in, it would be cramped. The real problem would arise once they got to orbit and had to move some of the storage bags to get access to the airlock. I could see why Dom was concerned, but it was hard to gauge how much of a problem it could be. Clutter tended to bother some people much more than others. Many astronauts had told me that without gravity, every cubic inch of volume could be used, making the problem much less significant than it appeared on the ground.

The next day, I called one of my friends who had commanded several missions to Mir and ISS. I asked him if we were getting close to a safety issue.

"I don't think so," he said. "The most critical events occur with all the stowage safely put away. The worst case would be an emergency descent due to a cabin leak or something like that, and then they would just reenter with the junk wherever it was. There isn't a second to waste in that situation, but it is very unlikely, and a few more bags of stuff won't make a big difference."

When Dom came by, I had Susan Creasy and Chirold Epp with me.

"I am very concerned that the stowage in the mid-deck is a safety hazard," he said. "It is fine while it is stowed, but as soon as we get to orbit, we'll have to pull things out, and the whole cabin will get extremely cluttered."

"I understand your concern," I said, "and I went over to see what your cabin configuration looks like. I also spoke to experienced commanders for their take, and they think we are okay. The problem is that I honestly don't know where the line should be. As you know, the issue belongs to the Space Shuttle program; they have reviewed it multiple times with safety and CB"—the astronaut office—"and all accept it."

"I know, but I think they are wrong. It puts us in a very difficult situation to deal with any problem that might come up."

"I hear you and respect your judgment, but I have no basis to challenge the Space Shuttle program. What I can tell you is that we have an enormous backlog of items that need to be flown. Some, like spares, are important to the safety of the station."

"But there are lots of other things in there too. Things that are not critical."

"That is true, too, though many of those things are at the request of the crew. We do the best we can with what the Space Shuttle program has given us. If you convince them to reduce the stowage, I won't fight you, and we will adjust our manifest. But as of now, I can use every bit of cargo space I can get."

Dom was clearly not impressed with our talk, but this was an issue with no obvious right or wrong answers. If there were an accident, people might criticize how much cargo had been carried, but for now, it was hard to know how much was too much.

My neutrality on this was partially shaped by the strong opinions coming from experts on a wide range of issues. Only a few weeks earlier, Frank had chided me over a decision I'd had to make about the hardware configuration on board.

The Russians had strict policies on how to manage backup and emergency systems. They took a hard line because some crews had not done what they'd been expected to do. The worst case had been discovered during the fire on Mir when Linenger had found the fire extinguishers still bolted in place.

On Mir, the crew had gotten away with a lot because the Russian comms had been so poor. But now, our access to satellite relays gave the ground nearly constant communications with the crew, including regular video. During one of these video periods, the Russians had noticed that the backup oxygen generator, the Solid Fuel Oxygen Generator (SFOG) was no longer installed in the *Zvezda*. Originally mounted on the wall above the table where the crew ate their meals, someone had removed and stowed it. The Russian ground team had wanted it reinstalled and ready for immediate use if necessary.

The crew had balked. They'd admitted that they had removed it but said it was an obstacle to their living space. They'd disobeyed a ground command, and since Frank was the commander, the Russians had expected me to order him to comply.

Accompanied by the flight director, I'd given him a call.

"The Russians want the SFOG back in place," I'd said. "It is their module, so I have promised to abide by their procedures."

"But we don't need it," Frank had objected, "and with our multiple sources of oxygen, we would have plenty of time to put it back if we did."

Frank had used a tone that scolded me as if I were still his deputy, but those days were long gone.

"I hear you, Frank, but this is an important matter of principle. We are talking about the Russian segment, and since I hold them accountable for its performance, I have to respect their requirements. In this case, their hazard controls assume that hardware is in place. You are complaining about convenience, not safety, so I have to go with their decision."

"It makes no sense!" he'd said.

"Frank, it's my job to lead the team, and I want you to put it back!"

He'd clearly been unhappy, and I am sure he'd thought I'd been intimidated by the Russians, but he'd been dead wrong. His responsibility during Phase 1 had been trivial compared to mine now, and I'd been disappointed that he had not appreciated that. He, of all people, should have understood that I'd had to stand up for our partners.

When I discussed the conversation with Missy, she was blunt.

"You are expecting too much from Frank. Remember all those times in Phase 1 when we had perfectly rational conversations with our crews before flight, and then they became myopic once they got to space?" she asked. "He is just doing the same thing. My guess is that everyone will."

She was right again. She and Frank got along well most of the time, but there were days during his mission when she was really angry at him. Still, she never asked for my help, and I stayed out of her way. She was very capable, and I had more than enough to keep me busy.

We did our best to take care of Frank, as we would any crewmember, but we were delighted to see how excited he was to be doing an EVA. As a shuttle commander, he had not been trained for EVA before coming to ISS, so this was a real treat for him. He would be in the Russian suit and operate out of the Russian airlock that arrived during his stay. It was a highlight of his career.

September 11, 2001, dawned bright and clear in Houston. It was a normal Tuesday, a busy day for the ISS crew with no shuttle present. After my normal rounds with the engineers and flight director, I chaired the MMT.

During the meeting, one of the flight directors came in and whispered in my ear.

"An airplane just hit the World Trade Center," he said.

Knowing that an airplane had once hit the Empire State Building in bad weather, I asked, "What is the weather up there?"

The flight director was taken aback by the question and said he would find out.

A few minutes later, he came back, looking grim.

"The plane was hijacked. We are securing the building."

Not knowing what to say, I nodded and continued the meeting. I could not get worried by a few words.

Wrapping up the MMT meeting, I left the telecon system up so that some of the specialists could discuss the details of the upcoming activities. I walked across the hall to the increment management center, where the television showed an image of one of the World Trade Center towers burning fiercely. As we watched, another plane swooped in and hit the second tower.

I froze in mid-step. Up until that moment, I had been sure that the first crash had been some kind of bizarre accident, but this was obviously not an accident. Someone had intentionally flown an airplane full of people into that building.

After a moment, I turned around and went back to the MMT room. Reclaiming my seat, I addressed the net.

"Victor, are you there?" I asked.

"Yes, Jim," he replied.

"Everyone, please listen carefully. The United States is under attack. I just watched an airplane intentionally crash into one of the World Trade Center towers in New York. The other tower had already been attacked, and we hear that other airplanes have been hijacked."

Pausing for a moment to gather my thoughts, I continued. "Victor, we are securing the control center here, and I give you my word that we will take care of the Russians and other internationals who are here with us. Please give me your word that you will do the same for the Americans working there with you."

Victor's voice was somber. "Jim, we will do as you ask. Please let us know if you need our help in any other way."

"Thank you, Victor. Let's keep everyone safe."

Leaving the room, I went down to the ISS control room. Randy Stone was there, talking to the flight director. He turned to me.

"Roy asked me to oversee our response to this attack." Roy Estess was the acting JSC director. "I have directed that we switch to the backup control center. We will secure this building except for a skeleton crew of volunteers needed to keep the main computer complex operating."

"How long will that take?"

"Our requirement is to be up and running in three hours, but I think we will do better than that. People are already configuring the consoles."

The backup control center was another facility on JSC that could perform all essential functions but was not publicly acknowledged. While I had thought that secrecy unnecessary in the past, now it made perfect sense.

While the team was beginning the transition, I turned my thoughts to the crew. Frank was the only American on board, and he needed to be told as soon as possible. I asked the flight director what had been done about it.

"The docs are getting ready to talk to him," they said. I had no need to get in the middle of that, but I did take a few minutes to talk to him when the docs were through.

While I was talking to my increment team, someone came through to ask us all to move our cars. Security was clearing the parking lot to make sure there was no car bomb that could take out the building.

Everything in the country seemed to come to a stop except for the disaster responses in New York and Washington. All our meetings were canceled, and by early afternoon, I was headed home. It felt strange to be going home in full daylight.

Overhead, there were no contrails, no airplanes coming or going. Doris was off on a trip to Huntsville, Alabama; she would drive home in her rental car.

36. THE END OF MY ERA

The weeks following 9/11 were confusing. Our work continued almost unabated, but everyone knew that something profound had happened. The conflicts of the world were no longer "over there" but in our own backyard.

A whole new set of concerns and risks seemed to explode onto the scene. A week after the 9/11 attacks, letters containing anthrax spores were sent to American media and senators. These letters were intentionally sent to incite anger and fear among the population—and they did.

With most of the anthrax attacks in Washington, DC, we in Houston felt somewhat removed from the risks. But our leadership would not allow us to become complacent. Massive new efforts were undertaken to protect the workforce and physical plant of NASA centers. Security was dramatically increased, with frequent vehicle searches, controlled parking near important buildings, and many other measures.

The shuttle and ISS received special attention. Everything intended for spaceflight underwent additional inspections and was held in secure storage to prevent tampering. This was already standard practice, but we refined an already-stellar process. NASA's astronauts and vehicles were viewed as high-priority targets, subject to special scrutiny and preventive measures.

The threat of a biological attack led to a requirement to fly antibiotics to the station to treat the crew should it be discovered after the fact that contaminated items had unknowingly been transported.

The prudence of that decision came into question when Susan Creasy called with news that the package of antibiotics was going to weigh five pounds.

"The docs are demanding we fly one thousand pills," she said. "That seems a bit much to me."

I had to agree, so I called Craig Fischer, the head of flight medicine, and asked for an explanation.

"That's the right number," he said. "We have seven people on the shuttle and three on the station. We need to be able to treat the shuttle crew until they get home and the ISS crew for a full course of treatment. Since not everyone tolerates the same antibiotic, when you run the numbers, you come up with five hundred tablets each of two different types. I know it's painful now, but the good news is that we can leave them on board and not fly them every time."

Craig and I had spent a lot of time together, and I trusted that his logic was solid. Although a review might have saved a pound or two, arguing over such a small amount would have been a waste of time when we had more important things to worry about.

The next shuttle flight was the one Dom Gorie was to command and the first of our utilization flights that directly supported the use of ISS for research. It focused on major experiment racks and logistics needed to make the station a real research facility. The flight marked a subtle shift in the program as it prepared to give more attention to the research community.

Adding research increased the demand for logistics, and a Progress vehicle was scheduled to dock the day before the shuttle was to launch. The timing was closer than we preferred, but practical schedule constraints left us with few options.

Progress vehicles had flown many missions, and it was easy to take them for granted. As if to warn us against complacency, a problem popped up with this one. After a normal rendezvous and docking, the latching hooks failed to close, preventing an airtight seal. This meant that although the two vehicles were joined, the crew could not open the hatch to access the logistics. Worse, the attachment might not have been able to bear structural loads from the shuttle docking or the station reboosting. The shuttle could not launch until the problem was resolved.

The Russians could not immediately tell us what had caused the problem or how they would fix it. We stood by while their engineers investigated. The next decision point for the shuttle launch was at the seven o'clock in the morning meeting to approve fueling the external tank, so I requested a report from the Russians by six o'clock.

Working through the night, the Russians believed they had identified the culprit. Camera views from the departure of the previous Progress showed what appeared to be one of its O-ring seals stuck on the docking system surface. The seal could have interfered with the final movements of the latching system. While this explanation was plausible, the Russians requested another day to be certain.

I agreed and began planning our options. I told Tom Martin, the MER manager on duty, what I wanted done so we would be ready to proceed once the Russians confirmed their findings.

"As you work through your normal process, map your logic on a decision tree that shows each potential action, why we think it's appropriate, and its consequences, including risks. If you have a recommendation, that's great, but what I am really asking you for is a logical structure to support our decision."

"We can do that. When do you want to see it?"

"We don't need an answer until tomorrow morning, but the Russians should be pretty mature with their analysis by the end of their day today. Because I want time to digest it, I'd like a meeting at four this afternoon to go over what you've put together. It may change overnight based on what the Russians find, but the logic should be good for whatever comes."

"And you want all of that by four? You don't ask for much, do you, Jim?" Tom was smiling when he said it, but I knew he was concerned about getting the team to agree on an approach by then. He also knew that I had been pushing decision trees to guide our choices for months.

That was the real point behind my request. We had successfully worked through hundreds of anomalies, but we sometimes wasted valuable time debating insignificant points. If the MER could lay out the logic of the decision early, everyone would understand the problem and its solution simultaneously.

"Remember, I'm not asking for a decision," I said. "That will come in the morning. What I want is the structure to drive the decision. If you do it right, everyone will see the correct answer as soon as the Russians give us the results of their investigation."

"We'll do our best!"

Upstairs in the control room, I asked the flight director whether the shuttle could dock with the Progress as it was.

"Nobody knows whether it would actually damage the hardware," he said, "but it could. That's why it's against the flight rules."

Although the MMT could waive the flight rules in an emergency, we were working hard to prevent that scenario.

I went down the hall to the Space Shuttle program's conference room where Jim Halsell was running the "tanking meeting." This was where the Space Shuttle program team confirmed that the hardware, weather, and other elements were ready so they could load the external tank with its fuel of liquid oxygen and hydrogen. The decision was taken seriously because every time the tanks were filled with these cryogenic liquids, it caused a little wear and tear on the hardware.

After reviewing the status of the shuttle and its support systems, Jim turned to me for an update on ISS. I outlined what we knew and what remained unknown, requesting another day to review the data. Even as I spoke, I was amused by our formality—there was no way the shuttle would launch unless the station was ready to receive it, yet my status report took the form of a recommendation.

After Jim accepted our recommendation and told the team to be ready to meet again in twenty-four hours, I mentioned that we would hold a telecon at four in the afternoon Houston time to review what we knew. The Space Shuttle program was welcome to listen, I said, but this was strictly an informational meeting—our formal decision would come in the morning.

After spending most of the day reviewing manifests and mission plans for future flights, I walked back to the Mission Control Center for the four o'clock telecon. After climbing the stairs to the second floor, I was surprised to find the hallway leading to the conference room packed with people eager to hear what was going to happen. Once inside, I made my way to the head of the table. Bill Gerstenmaier and Tommy Holloway were both present, though they sat back from the table, clearly leaving the meeting to me.

When the meeting facilitator handed me a list of attendees on the telecon, it seemed as though the whole of NASA was there, including center directors and many senior officials from headquarters. After introducing the purpose of the meeting, I turned it over to Tom Martin. He began by explaining how this situation differed from what they would usually present to the MMT. Instead of presenting data on potential failures, which was all on the Russian side—he would outline the logic behind our possible responses to the problem. He delivered a complex argument that spanned multiple charts, but the logic and conclusions were thorough and well-structured.

Although he was careful not to draw any conclusions, the facts suggested that the Russians would conclude that an EVA was going to be required to remove the stuck seal. While this seemed likely, it was not yet confirmed.

"Jim, this is Ron." The voice of Space Shuttle Program Manager Ron Dittemore came through the phone. "I appreciate the thorough discussion the MER team put together—the logic is very helpful for those of us who don't study the Russian systems. But since everyone agrees that an EVA is going to be required, it seems to me that we can go ahead and make the decision tonight to delay the shuttle again."

Once again, and despite my best efforts, the Space Shuttle program was jumping into my process without fully understanding the implications. My entire program was built on hard-won trust and respect, and allowing the Space Shuttle program to get ahead of the Russians would not be helpful.

Before I could respond, Gerstenmaier moved up to the table, indicating he wanted to say something.

"Ron, this is Bill. I agree that it seems likely that an EVA will be required, but we don't know that for sure. The Russians may come up with another solution, and it's important that we give them the time they need to go through their process and make their recommendation to us. We don't want to short-circuit their role."

"I hear what you're saying, Bill, but I also don't want to waste our time when we already know the right answer."

I punched the mic button hard and spoke calmly but firmly.

"Ron, we don't know the right answer! We only know the basics of their system, and we would be foolish to assume otherwise. We already have experience with Americans thinking we knew the Russian hardware better than the designers themselves—it has cost us dearly. That's precisely why I didn't ask the MER to make a recommendation. Their task was to structure the problem in a way that would allow us to make the best possible decision when the time came. Right now, we do not have the necessary data, and it would be harmful to the partnership to pretend that we do."

"Okay, Jim, we'll wait for the tanking meeting." Ron sounded like he was laughing, humoring me.

As the meeting wrapped up, the crowd quickly dispersed, leaving Tommy, Bill, and our core ops team in the room.

"Any word from the Russians?" Tommy asked.

"Not officially, but our folks in Moscow are pretty confident an EVA is going to be required. The big issue is that it involves reaching into the area between the Progress and station hatches. We want to make sure that the crew doesn't have to put an arm into harm's way."

Nodding, Tommy got up and left, followed by Bill.

The next morning, we got a crisp reply from Blagov—an EVA was indeed required, with about five days needed for planning and execution, exactly as our folks had anticipated through the decision tree. I called Bill and asked him to pass the update to Tommy.

At the tanking meeting, I briefed everyone on the ISS decision. Halsell had attended the meeting yesterday and was ready with directions to reset the launch attempt for the day of the EVA. As soon as we knew that the EVA had been successful, they would proceed with the launch.

The following days passed in a blur as the Russians put together a plan to remove the seal, designed a tool the crew could fabricate to keep their arms out of the mechanism, and briefed the crew. Their process even included putting the crew through a dry run. NASA followed along and concurred on every step. Frank Culbertson helped his crewmates prepare for the EVA and supported them by managing the station systems while they were outside. It was exactly the kind of teamwork I had been working toward since the early days of Phase 1.

The EVA went perfectly. The Progress mechanism was extended to give access to the docking surface, and the crew used the tool to reach in and remove the seal. The docking mechanism was retracted and sealed. I immediately called Halsell to confirm we were ready for the shuttle, and they prepared to launch. Ultimately, they had to slip one more day due to weather at the launch site, but overall, our response to the problem had been remarkably smooth and efficient. I was proud of the team and the professionalism everyone displayed.

The next day, we held a telecon with the Russians to discuss the work and their findings. Although they could not figure out why the seal had come loose, they did confirm it had come from the previous Progress. Blagov even promised to send me a piece of the seal as a memento. That piece is now mounted and displayed on the wall of my home office.

When the shuttle finally launched, for some reason, I didn't feel engaged. I had a distressing feeling that I was watching the mission on TV rather than participating.

When the shuttle docked and the joint mission operations truly began, I finally realized that this was the sign I had been looking for. Instead of feeling in charge and responsible for everything that was going on, I felt disconnected. Everything was running smoothly, and the team was doing great, so I waited until the next day to talk to Tommy.

"I am cooked," I said. "The pace of the last eight years has caught up with me, and I just can't do the job as well as it deserves. The time has come to step down."

Tommy acted as if he had been expecting this for some time.

"Can you get through this flight?"

"Absolutely. If a problem comes up, the adrenaline will kick in, and I will be back in the middle of everything. Heck, I could probably go on for another six months, but I am worried that the quality of my decision-making would not be up to snuff. You need someone who is fully engaged and not burned out."

"Okay, we'll talk about it after undocking."

A few days later, when the shuttle was on its way home, it was announced that Mike Suffredini would be taking my place as chair of the MMT. He had been running the engineering team responsible for the US elements, so he was up to speed on the NASA hardware. He had also begun to learn the Russian systems as we'd worked through their problems. His biggest liability was that he was not known or trusted by the Russians. His rather caustic personality was also an issue, but he knew he would have to overcome that in this new role, and he made a conscious effort to do so.

I helped him through his first two MMT meetings, during which he did just fine. Although he had rarely attended my MMT meetings in person, he'd listened regularly to keep on top of what was going on. I also knew he had been making sure that I'd taken care of "his" hardware. The holidays in the US and Russia allowed a bit of a lull, giving people time to catch up without high-risk operations. I offered my services for anything that might come up but soon stopped coming to the meetings. It was time to unplug.

Roy Estess, the acting center director, supported my decision to step down. Although Roy was the permanent director of the Stennis Space Center, he had been called in to take George Abbey's place after Dan Goldin had removed George almost a year earlier. Unlike George, Roy made an effort to reach out to all the people who worked for him. He had paid particular attention to the operations of the ISS team, and I was both surprised and relieved that he was so familiar with my work. He apparently had watched my work more closely and appreciated the challenges and the workload I had endured better than Holloway ever had.

"You've gone far above and beyond the call, Jim," he told me, "and we are all so very grateful for your extraordinary efforts. Even in the short time I've been here, I can see the effects of the stress on you. You need to take some time to unwind, then we'll find you a job as a director here at the center. You deserve it, and we need your experience and smarts. In the meantime, I'll make you a special assistant so you can stay engaged as much as you want."

Randy Stone, who had become Roy's deputy, echoed Roy's support. "When you're ready, we have lots of things that need your attention. You have earned our eternal respect and gratitude."

It was nice to know my efforts had been noticed.

Tommy remained the exception. He seemed almost relieved that I was finally going to be out of his hair. When the ISS team threw a going-away party for me, he came to make the official presentation, as though going through the motions. But as he read the synopsis of my work that someone had prepared for him, he was obviously startled by what he read.

"It says here that you created the architecture for the program and the vehicle! I had no idea you had touched those so profoundly. It also says you have been responsible for thirty-six American EVAs on your watch. NASA has only done 104 EVAs total!" He went on to quote other accomplishments, frequently looking up from the paper to me, obviously ignorant of the challenges I had overcome.

Others from the team took turns getting up to say the usual nice things that people say about a departing boss or colleague. Many of them described me as incredibly patient and tolerant of the complaints and problems that never stopped.

At this, my wife spoke up. "Okay, hold on! Let me get this straight! We are talking about my husband here, right? Jim Van Laak?" She was clearly teasing, and the room shared the laugh, but everyone understood that my time on the job had cost both of us dearly.

When it was time to say something myself, I was too emotionally drained to say much, but what came out was something like this:

"I hope you all know what an honor it has been to have worked with all of you on something truly amazing. Words cannot express my sincere appreciation for your patience, help, and support. Each and every one of you did all I could ask and more. It has been a once-in-a-lifetime experience and privilege to lead such a spectacular team.

"And more than I can possibly say, I thank my wonderful wife and family for tolerating the incredibly few hours I had for them and for forgiving my impatience and sour mood.

"Many have pointed out that flying a shuttle flight is a sprint while the station is a marathon. But for me, it has been a relay race. The job took everything I had for as long as I could give it, and now it is time to pass the baton."

Looking at those who had stood by me the most—Randy Stone, Susan Creasy, and Missy Gard—I went on.

"A few of you know that it took everything I had to hold this program together. Some of you will also remember when it looked like a fool's errand, certain to end in tears. But with your help and more than a little luck, we survived the most dangerous times. I am proud to turn over responsibility for this fine ship to this great team. For as long as

I live, I will remember these as the most amazing and rewarding days of my life.

"Thank you for sharing them with me."

There were lots of warm handshakes and hugs, softening the pain of leaving this extraordinary job.

37. WHAT NOW?

The year 2002 dawned without ISS in my life. For the first time in over a decade, I did not have people waiting for me. Although I tried to luxuriate in the freedom this offered, I mostly felt empty. While I missed the program intensely, I was more certain than ever that my decision to leave had been correct. Early in January, I met with Roy Estess about my future.

"I want to make you a director, and I have a couple of options. But the truth is that O'Keefe"—O'Keefe was the NASA administrator named by President Bush—"is about to name my replacement, and it would be better if he selected you for your new job. Just be patient, and good things will come. You've earned it!"

Although there were a few rumors about who the new director might be, including some who were close friends, I had a sense of foreboding that my future would be in the hands of an outsider.

The prospect of leaving JSC did not cross my mind. After twelve intense years there, I could not imagine working anywhere else. The professionalism, work ethic, and intense commitment to the mission resonated with something deep within me. This special group of people had become close friends through the most intense experience imaginable, short of a war.

A few weeks later, retired Marine Lt. Gen. Jefferson Davis Howell Jr. was named the JSC director. I had absolutely no idea who he was, and most of my colleagues knew little more. This was odd, as we soon discovered that he had been a VP in charge of the contractor supporting the JSC safety office for some time. Somehow, despite that significant role, he had managed to remain almost completely unknown.

Our new boss went by the nickname "Beak," his marine fighter pilot callsign, which came from his rather large nose. He wanted it used under all but the most formal of circumstances.

Having been a fighter pilot myself and around them since I was a young man, I felt perfectly happy to accept him as another member of the club. Most of them were sharp and responsible professionals; it takes a great deal of hard work to get the most out of such sophisticated machines.

Beak quickly made it clear that he was proud to be "just a dumb fighter pilot." Although I expected this self-deprecating expression as part of the macho fighter pilot image, I soon learned that it was closer to the truth than I was used to.

Our first meeting went well, and he welcomed me as a member of his senior staff. When I told him of my background in the air force, he responded with some good-natured kidding about the relative manhood of marines versus air force pilots. But while he seemed friendly enough, I was bothered by the casual way he approached our business. Most NASA professionals take their work very seriously.

A few weeks after his arrival, I was attending one of his senior staff meetings, during which they briefed the boss on issues being worked through by the major programs. That week, the issue was a concern about a particular solid rocket motor. Ron Dittemore was discussing the issue when Beak broke in to ask a question.

"How do you shut off the solid rockets if you have a problem?" he asked.

You could have heard a pin drop as the staff struggled to maintain their composure. One of the most elementary facts of rocketry was that you could not shut off solid-fuel rockets. It was deeply disconcerting to find that our new boss was so ignorant of the rudiments of our business, especially since he had spent several years running our safety support contractor.

After a moment, Ron replied. "One of the disadvantages of using solid rocket motors is that they cannot be shut down. Once you light them, they burn until their fuel is gone—about two minutes."

Beak nodded, and the meeting went on. After the group had left, I stuck my head into his office.

"Beak, I usually know what is coming up in any meeting on your schedule. If you'd like, I would be happy to give you a quick briefing on the issue so you can avoid awkward questions."

"Thanks," he said jovially, "but I like to take 'em on the chin!"

I did not know exactly what that meant but concluded that he did not want my help. This became clearer with each succeeding interaction. He was ignorant of our work and the technical issues underlying it and apparently content to stay that way. He was more interested in political issues, like getting funding for a new life science lab, than safety issues on upcoming missions. This was distressing since the center director had a key role in setting the tone for safety in everything we did.

My view of the work at the center was also changing. Since I was no longer running the ISS MMT, I had time to stand back and take a broader look at what was going on in our programs. The ISS work was proceeding pretty much as expected, but there were subtle changes on the Space Shuttle program side. Some of them had been faintly visible to me even from within ISS, but now they were becoming clearer.

The biggest issue was the increasing emphasis on the cost and schedule of our work, especially the date when the ISS would be completed. Maintaining cost and schedule targets is always a challenge, but the annual budgets for ISS and the Space Shuttle program were pretty much set in stone. This meant the only flexibility we had to respond to problems was to slip the schedule. Even so, with the exception of the initial delays in completing the *Destiny* lab, our schedule performance had been outstanding. Given the huge risks and multibillion-dollar assets we managed, the new focus on launch dates seemed foolish and potentially dangerous.

George W. Bush was touted as the "MBA president," and our headquarters team was telling us that he was cracking the whip about work getting done on time and on schedule. NASA Administrator Sean O'Keefe was closely monitoring the planned schedule and making noise when things slipped. In the past, the idea of letting the completion date slip a little was seen as a small price to ensure safety. No more. O'Keefe demanded a firm schedule to complete the ISS assembly. While I never heard anybody say safety was unimportant, there was a palpable indifference to the issues we faced. Like Beak, O'Keefe lacked a technical background and was not interested in the reasons for delays. His demand for schedule discipline was felt keenly throughout the system. When I visited a friend at NASA Headquarters, he pointed to the screensaver that had been installed on his computer. It counted down the days, hours, minutes, and seconds until the ISS assembly was to be completed.

"It's on every machine in the building," he said, "and it comes up in almost every conversation."

To many of us, this was a sign that the organization had lost its safety focus. Back at JSC, I kept alert for signs that this damaging attitude was affecting the working troops. At Space Shuttle program meetings, I started to see what I perceived as a lessening of the rigor applied to anomalies that arose during missions. If true, this would mean an increased risk. I never found an obvious smoking gun, but I saw growing evidence that schedule urgency was overtaking a safety-focused process.

One day, when I was talking to Beak, I felt the time was right to express my concern to him. Throughout my career at JSC, my bosses had always wanted to hear about any safety concerns, no matter how remote. It seemed natural to treat him the same way.

"Boss, I am becoming concerned. I think I see schedule pressure making the Space Shuttle program truncate anomaly resolutions. If I am right, we are ripe for an accident!"

Beak's response was silence. He quickly ended the meeting and said nothing more on the subject. Two days later, he came to my office.

"You are not a team player," he said, "and you need to find another place to work!"

Utterly stunned, I did not know what to say.

He went on. "I know what that feels like, like somebody just kicked you in the gut. But you will find another place, and it will be better for you."

With that, he turned around and walked out, marking the last time he would ever speak to me.

> **Commercialization**
>
> In this time frame, some people thought NASA should turn over the reins of the Space Shuttle program to a private company. Those who favored this view argued that NASA was too conservative and "gold plated" every aspect of its work. They believed that the technology and operations had matured sufficiently to allow NASA to hand it over to a company focused on efficiency. This was supposed to allow NASA to save money and focus on the next step in space exploration.

The drive home was empty. It was incredible to believe that after all that had happened, after all that I had accomplished, I was suddenly out. The next day, I talked to Randy Stone about it, but he was sheepishly quiet. He was uneasy about his own position as Beak's deputy, and I could tell he was embarrassed about what had happened and his own inability to change it.

Over the next few weeks, I tried to make sense of this development and what it meant for the future. As much as I loved working at JSC, I was not eager to go back to the ISS program, and I was not even sure Tommy would have me. For sure, I did not want to work under Ron Dittemore in the Space Shuttle program.

Wondering what I had said or done to elicit this response from Beak, the most likely answer was that, like many general officers, Beak was used to subordinates who saluted and obeyed without question.

Instead, I had told him what I thought, truthfully and directly, and he hadn't liked it.

The realization was terrifying. The very thing that had drawn me to the space program was the overpowering intellectual honesty that it demanded. Although it was not always perfectly done, the pursuit of honesty was the core principle driving every person I admired and respected. Our work was so demanding and the risks so great that we were ethically obligated to give our honest opinions without regard to how well they would be liked.

I harkened back to the evening when Kranz had told me the story about the Apollo fire and the creation of the MOD motto, "Tough and Competent." His words and example had inspired me like none other, and I had always done my best to emulate his example. This had fueled my success at JSC, and now it was my undoing.

A wave of nausea surged through me as if my world had been turned upside down. If NASA management only wanted yes-men and would punish those who spoke truth to power, the integrity of our processes would soon evaporate. Even well-meaning people would see the shift and would subtly adjust their behavior to fit the new circumstances. Slowly, often unconsciously, safety and mission success would suffer. It was obvious that while Beak talked about safety, his words were hollow, without understanding or integrity.

It was only logical to wonder about the rest of the NASA management team. Sean O'Keefe seemed likable, but his words and actions put him on a collision course with our mission. Had he been humble and wise, he would have found that most decisions were sophisticated technical judgments, not casual choices, and relied on the judgments of his technical experts. Instead, he was cocky and aloof, and many of his decisions showed profound ignorance.

At least, I thought so, but there was no way to be certain. Like so much in life, this was an exercise in judgment and highly subjective. Who was I to insist that my subjective judgments were better than those of others?

I thought back on the foundations of our success. Every mission had been analyzed beforehand to predict the behavior of the systems and the margins we had before failure. After the mission, we would compare that predicted performance with what we'd actually seen to find anomalies beyond our experience. Even after one hundred shuttle missions, we were still seeing enough anomalies to know that we were operating at the edge of our capabilities and that there were things we really did not understand. That knowledge should have fed profound humility.

If I was right, that humility was fading, and with it the caution that had kept us safe. My comment to Beak had been my way of saying that I thought NASA management was becoming complacent, leading to increased risk.

Seeking another perspective, I shared my concerns and Beak's response with Michael Greenfield. He was unimpressed with both my concerns about safety and my mistrust of Beak. This was odd because Michael was both deputy associate administrator for safety and keenly aware of Beak's lack of technical depth.

Not satisfied, I mentioned my concerns to Bill Readdy, a former astronaut and head of the Office of Human Spaceflight at headquarters. Bill's response was terse: "Well, another party heard from."

He, too, showed no interest in the source of my concerns or my advice on how to address them.

This cool response from people I respected caused me to reassess my judgment. I wondered if I was making too much of the problem. My judgment that the Space Shuttle program was cutting corners was based on my reading of the events, along with my life experiences, values, and opinions. While I had done well in the past, I knew I could be wrong. It was hard to ignore the fact that many people I respected disagreed with me.

After much thought, I decided that I had done my duty by expressing my concerns to agency leaders. If my views had merit, they should see evidence that would validate them, hopefully well before an accident would occur.

A few months later, on shuttle flight STS-112, a chunk of foam separated from the external tank and hit the attachment ring for one of the solid rocket boosters. While this itself was not alarming, the chunk was much larger than our previous experience. If the anomaly resolution process had been working as intended, it would have been investigated thoroughly, but that might have delayed the next launch.

Instead, Ron Dittemore made the event an "action item" that did not require a formal investigation. A significant event that should have been thoroughly investigated instead disappeared into the river of routine work.

Two flights later, on STS-107, shuttle *Columbia* lost another large chunk of foam from the external tank, but this one hit the left wing of the orbiter. Because one of the tracking cameras was not working that day, there was no way to determine where on the wing it had hit. Aerodynamic models said that debris could not hit the leading edge, which was the only critically vulnerable part of the wing.

Throughout the flight, engineering and management discussed what, if anything, to do about the hit. Later, critics would make a great deal of the fact that the model used to assess damage to the tiles on the underside of the wing had not been very accurate, but in fact it had not mattered because the tiles could have tolerated enormous abuse and still done their job. The worst-case scenario seemed to be that the wing could take a long time to repair during ground processing.

History now knows that those aerodynamic models were wrong. The piece of foam that came off was large and flat, generating lift that caused it to move far enough to hit the wing's leading edge. It opened a hole that allowed super-heated plasma into the wing during entry, leading to the wing's structural failure. *Columbia* and its crew disintegrated as they reentered Earth's atmosphere. I lost four good friends, including a neighbor, in the accident.

Without question, mistakes had been made, but many of NASA's critics have been dishonest in their criticisms. Even the *Columbia* Accident Investigation Board did not fully understand the NASA process and culture. The accident resulted from two important errors, both made long before the fateful launch. The first was in the aerodynamic model; the second was management's decision to cut short the anomaly resolution process. The mistakes made by the shuttle team during the mission were predicated on those two mistakes, not rampant carelessness by managers.

The loss of *Columbia* was an unmitigated disaster. Public support for NASA was deeply damaged by irresponsible statements made by those who should have known better. Though well-meaning, they let their opinions speak rather than the facts.

The loss of *Columbia* might have been the death knell for the International Space Station had it not been for the decision to include the Russians. Their Soyuz and Progress crafts would continue to bring up cargo and rotate crews so the ISS could remain occupied, productive, and safe while the Space Shuttle program team worked on returning to flight. Without them, NASA would have been forced to either accept high risk to return the shuttle to flight before the investigation was complete or allow the ISS to reenter and burn up in the atmosphere. Our pioneering outpost survives and thrives only because it was a truly international endeavor.

Healing Together

Columbia was lost on a Saturday. The following Monday, a group of teachers gathered at Space Center Houston for a previously scheduled conference about using ISS to motivate STEM students. Since I was not part of the accident investigation, I was asked to talk to them. With no prepared thoughts or briefing material and barely in control of my own emotions, I took the stage in front of hundreds of tearful teachers. Speaking quietly and honestly, I told them about our space program, its challenges, its heroes, and why it was all worthwhile. While I hope I helped them through their ordeal, I know their attention and concern helped me through mine.

At the time of the accident, I'd been focused on finding a new job. The tragedy left me deeply conflicted, wanting to be helpful in the recovery from the accident but sickened by the leaders who had caused it and those who exploited the outcome for their own benefit. I had a profound sense that my time at JSC was over and that it was time to just move on.

In June, I left JSC for a job at the NASA Langley Research Center. On one of the most difficult days of my life, I fought back tears as I turned in my JSC badge. Whatever lay ahead, I knew I would never again work with a group of people with such astounding energy and intellect.

AFTERWORD

June 7, 2022

Hampton, Virginia, lies quiet and green outside my window. Its brilliant blue sky and sweet breeze bear little resemblance to the frigid black of a Maine winter's night or the steamy heat of a Houston day. Its peace nurtured the healing required to finish this book, to tell this story that needed to be told and to continue my work on the space program.

After eighteen years away for the manned spaceflight program, I once again found myself working on the frontier of humanity's expansion into space. Since March, I had been consulting with the Johnson Space Center's team on the Artemis program, helping them work through the challenges of returning humans to the moon. I had become the gray-haired man, filling the role that people like John Aaron had played for me during my time in the hot seat on ISS. It felt like home.

Many of the people and organizations were familiar. Some who had worked for me early in their careers had achieved positions of leadership, though most of my peers had retired or moved on. Yet, as friendly and familiar as this was, I was surprised at how little the team knew about how others, including ISS, had solved similar problems—logistics, reliability, integration, and especially risk. While the challenges presented by the lunar pole environment were new and daunting, I sought to show them concepts and techniques that could save Artemis time, money, and loss.

Like ISS, their most daunting challenges were rooted in human rather than technical issues. Where I had faced a cultural clash with Cold War adversaries, they needed to make peace with a plethora of new commercial companies. As in ISS, both sides ardently believed that their way of going to space was the only right way and resented the need to work with others who were less enlightened. Once again, arrogance and ignorance limited the progress these amazing teams could achieve.

Perhaps the biggest difference for me was that I joined their Herculean effort from the quiet isolation of my home in southeastern Virginia rather than the bustling camaraderie of the steamy Houston campus. My office was a small, gray laptop on a folding table in an unused bedroom, and it was there that I received an email from Missy Gard telling me of the death of Valery Ryumin. The news caught me by surprise, as Valery always struck me as a man who would live forever—a timeless pillar of our shared ordeal.

Ryumin had been one of my closest Russian colleagues, sometimes partner, sometimes opponent, as we'd pursued our grand experiment in international cooperation. Our interactions, whether in the chill of a Moscow winter or the steam of a Florida summer, now seemed light-years distant from my quiet roost. Ours had been a difficult relationship, born during the turbulence that had created the ISS, tempered during the crises on Mir, finally culminating in the mutual respect that had propelled the ISS these past twenty years.

When I told my wife about Ryumin's passing, she remarked on the irony that we learned of it on the one hundredth day of the Russian invasion of Ukraine. Russian President Putin's idiotic war had largely destroyed the hard-earned progress our nations had made in overcoming the animosities of the Cold War, not only in space but also in economic and diplomatic spheres. Those of us who'd made the difficult transition from Cold Warriors to respectful and cooperative partners could only grieve for the damage one man's nineteenth-century values had done to twenty-first-century progress.

Because of my work with Ryumin and his team, I was not surprised that the great exodus of disgusted Russians since the Ukraine invasion had been preponderantly young, technical professionals. They and we know that the problems we must solve respect only the laws of physics, not political or philosophical differences. One cannot will a rocket to fly or decree that a computer must work. Our work is guided by the absolute truths of our universe, and it is only there that we can find our success.

Engineers also viscerally understand, far beyond philosophy, the power of cooperation: No great engineering enterprise can be accomplished without it. Engineers seek to align on a common purpose, adding their expertise constructively because it increases the power available to succeed. They know that anything that breaks their common purpose, from wars to bureaucratic infighting, consumes precious resources and makes progress much more difficult.

Ryumin and I confronted that reality directly and ultimately found a path to success. So long as the Russian flight control team refused to accept us into their work, both our programs suffered. Our research program suffered first, as our plans were decimated by unexpected interruptions, wasting precious time and money invested in scientific goals. But the far more serious damage came when Russia refused to accept our aid to deal with their problems driven by financial and human stresses. Only the existential crisis that followed the collision of the Progress cracked their stubborn pride and allowed us to contribute.

And contribute we did. Over many months, Frank Culbertson and I collaborated with Ryumin and his team to help them manage their many challenges. We flew hardware on the shuttle, offered the help of our crews, sought common solutions, and stood by them in the court of public opinion. It was a difficult time, with endless, grueling hours of working together and setting aside our pride, parochialism, and egos to find compromise and drive toward our common goal.

Of course, we were not alone. Thousands of others had to learn that same lesson, overcome their own fears and prejudices, and focus their energies on a cooperative future. It was difficult, and not everyone succeeded, but enough did to make it work. I remain enormously proud of the people who put our collective future above their individual needs, and I cherish the opportunity I was given to lead their work.

It was literally a life-changing experience. My efforts in building the International Space Station taught me more than any university or philosophy ever could. Its profound challenges and crushing responsibility gave me the perspective and knowledge to face any difficulty that the future may provide. They produced a long view of human destiny that reduces any challenge to its essentials. The result is a certain knowledge of how we can solve the many great challenges that must certainly lie ahead, from climate change to economic inequality to artificial intelligence to human rights.

Engineers live to tackle technical challenges, but these issues also come burdened with very human dimensions. Their political, philosophical, and even religious overlays can make cooperation difficult, but my experience with ISS shows that we can get past them if we focus on our common goals rather than our differences.

There were those among us who had the wisdom and foresight to see this, and they drove us toward this goal without regard to the barriers that lay ahead. They knew that those with the power to destroy humanity must learn to cooperate, or all would be lost. But many others, including me, were blinded by the gloom of our long Cold War night and slow to embrace this vision. Fortunately, driven to protect our precious space program, I and others came to see the vision as well. It is the foundation on which the ISS "stands."

NASA did not go into space for this particular long view any more than we went to the moon in the 1960s for the long view of our common humanity; the vision came unbidden. Space exploration has given us the opportunity and the motivation to embrace the challenge of working together on our precious home world. All that it demands is that we will it.

Afterword

Although the Russian invasion of Ukraine has cast a dark shadow over the world, ISS remains a beacon of hope for humanity's future, both metaphorically and literally. Its brilliant light provides an inspiration to continue, a reason to hope, and a prototype for the society we want for our children.

ACKNOWLEDGMENTS

This book and the events it describes would have been impossible without the generous support of many people.

My first thanks go to my parents and siblings, whose love and patience gave me a good home, a loving childhood, and a solid start to life. My dear mother played a special part, overcoming her motherly fears to support my love for aviation that helped me find myself.

Special thanks also go to my best friend from childhood, Rich Roberts, and his older brother Paul. They invited me to join their Project Goodwill, which mimicked Apollo with large model rockets, and let me touch the distant glory of the space program.

Aviation and aerospace provided the challenge I sought, the confidence I lacked, and the mentors who would guide my way. The first mentor was Waterman Farnum Brown VII, dear Brownie. His lessons literally saved my life and enriched it enormously.

The story touches on Michael Greenfield and Mike Smith, but neither gets the credit he deserves. Without their friendship and support, my participation in the space program would have remained a distant dream.

Dennis Webb gave me my first job at the Johnson Space Center in my beloved Mission Operations Directorate, along with the freedom and support to do what I thought best.

Rick Nygren stepped up when the chips were down, saving Shuttle-Mir from disaster. He found a way to work with the Russians despite the lack of support from senior leaders. He and his team of coura-geous youngsters preserved our program in its moment of need.

Special recognition goes to Randy Stone. His friendship, advice, and support sustained me; his professionalism and selflessness inspired me.

Two special Russians also helped me through these times. Victor Blagov and Valery Ryumin, raised under Communism and taught to despise Americans, overcame that history to work with us. They did what needed to be done.

Finally, none of my work would have been possible without the boundless support and inspiration of my family, especially my wife, Doris Hamill. She is my soulmate, woven through every fiber of my being and every day of my life. She paid a great price to love and sup-port me through the events in this book, for which I am profoundly grateful.

ABOUT THE AUTHOR

Jim Van Laak grew up in the old industrial city of Schenectady, New York. During the Apollo program of the 1960s, he became enthralled by the space program that dominated the world stage. Entranced by aviation, he got his pilot's license at age seventeen and soon thereafter survived a mechanical failure–induced accident. Inspired to learn what he could about these machines, he worked for an airplane mechanic while also attending college at Rensselaer Polytechnic Institute. A low draft number led him to join Air Force ROTC and later become a pilot and maintenance officer in the 49th Fighter Interceptor Squadron.

After leaving the air force, his deep knowledge of aircraft systems won him a role at NASA Headquarters, where he expanded his expertise to space shuttle systems. He then transferred to the Johnson Space Center to lead maintenance and logistics planning for Space Station Freedom. Having found a weakness in Freedom, he was tasked to ensure the International Space Station (ISS) program would not suffer the same vulnerability.

For ISS, he helped integrate the US and Russian space programs, creating a comprehensive risk management process to ensure success. Later, he became deputy director of the Shuttle-Mir program that sent American astronauts to Russia's aging Mir space station. His response to critical events developed the expertise and respect needed to manage spaceflight operations during ISS's critical start-up phase. He became the focus of conflicts between the US and Russian approaches to spaceflight, as well as between the existing Space Shuttle and the emerging space station programs. This allowed him to guide the programs through several critical tests of cooperation that could have destroyed the program.

Van Laak later became a senior manager at NASA's Langley Research Center, served as the FAA's deputy associate administrator for commercial space, and assisted in the launch and activation of the James Webb Space Telescope.

INDEX

9 781966 786580